Research Notes in Mathematics

Main Editors
A. Jeffrey, University of Newcastle-upon-Tyne
R. G. Douglas, State University of New York at Stony Brook

Editorial Board
F. F. Bonsall, University of Edinburgh
H. Brezis, Université de Paris
G. Fichera, Università di Roma
R. P. Gilbert, University of Delaware
K. Kirchgässner, Universität Stuttgart
R. E. Meyer, University of Wisconsin-Madison
J. Nitsche, Universität Freiburg
L. E. Payne, Cornell University
G. F. Roach, University of Strathclyde
I. N. Stewart, University of Warwick
S. J. Taylor, University of Liverpool

Submission of proposals for consideration
Suggestions for publication, in the form of outlines and representative samples, are invited by the editorial board for assessment. Intending authors should contact either the main editor or another member of the editorial board, citing the relevant AMS subject classifications. Refereeing is by members of the board and other mathematical authorities in the topic concerned, located throughout the world.

Preparation of accepted manuscripts
On acceptance of a proposal, the publisher will supply full instructions for the preparation of manuscripts in a form suitable for direct photo-lithographic reproduction. Specially printed grid sheets are provided and a contribution is offered by the publisher towards the cost of typing.

Illustrations should be prepared by the authors, ready for direct reproduction without further improvement. The use of hand-drawn symbols should be avoided wherever possible, in order to maintain maximum clarity of the text.

The publisher will be pleased to give any guidance necessary during the preparation of a typescript, and will be happy to answer any queries.

Important note
In order to avoid later retyping, intending authors are strongly urged not to begin final preparation of a typescript before receiving the publisher's guidelines and special paper. In this way it is hoped to preserve the uniform appearance of the series.

Titles in this series

1. Improperly posed boundary value problems
   A Carasso and A P Stone
2. Lie algebras generated by finite dimensional ideals
   I N Stewart
3. Bifurcation problems in nonlinear elasticity
   R W Dickey
4. Partial differential equations in the complex domain
   D L Colton
5. Quasilinear hyperbolic systems and waves
   A Jeffrey
6. Solution of boundary value problems by the method of integral operators
   D L Colton
7. Taylor expansions and catastrophes
   T Poston and I N Stewart
8. Function theoretic methods in differential equations
   R P Gilbert and R J Weinacht
9. Differential topology with a view to applications
   D R J Chillingworth
10. Characteristic classes of foliations
    H V Pittie
11. Stochastic integration and generalized martingales
    A U Kussmaul
12. Zeta-functions: An introduction to algebraic geometry
    A D Thomas
13. Explicit *a priori* inequalities with applications to boundary value problems
    V G Sigillito
14. Nonlinear diffusion
    W E Fitzgibbon III and H F Walker
15. Unsolved problems concerning lattice points
    J Hammer
16. Edge-colourings of graphs
    S Fiorini and R J Wilson
17. Nonlinear analysis and mechanics: Heriot-Watt Symposium Volume I
    R J Knops
18. Actions of fine abelian groups
    C Kosniowski
19. Closed graph theorems and webbed spaces
    M De Wilde
20. Singular perturbation techniques applied to integro-differential equations
    H Grabmüller
21. Retarded functional differential equations: A global point of view
    S E A Mohammed
22. Multiparameter spectral theory in Hilbert space
    B D Sleeman
24. Mathematical modelling techniques
    R Aris
25. Singular points of smooth mappings
    C G Gibson
26. Nonlinear evolution equations solvable by the spectral transform
    F Calogero
27. Nonlinear analysis and mechanics: Heriot-Watt Symposium Volume II
    R J Knops
28. Constructive functional analysis
    D S Bridges
29. Elongational flows: Aspects of the behaviour of model elasticoviscous fluids
    C J S Petrie
30. Nonlinear analysis and mechanics: Heriot-Watt Symposium Volume III
    R J Knops
31. Fractional calculus and integral transforms of generalized functions
    A C McBride
32. Complex manifold techniques in theoretical physics
    D E Lerner and P D Sommers
33. Hilbert's third problem: scissors congruence
    C-H Sah
34. Graph theory and combinatorics
    R J Wilson
35. The Tricomi equation with applications to the theory of plane transonic flow
    A R Manwell
36. Abstract differential equations
    S D Zaidman
37. Advances in twistor theory
    L P Hughston and R S Ward
38. Operator theory and functional analysis
    I Erdelyi
39. Nonlinear analysis and mechanics: Heriot-Watt Symposium Volume IV
    R J Knops
40. Singular systems of differential equations
    S L Campbell
41. N-dimensional crystallography
    R L E Schwarzenberger
42. Nonlinear partial differential equations in physical problems
    D Graffi
43. Shifts and periodicity for right invertible operators
    D Przeworska-Rolewicz
44. Rings with chain conditions
    A W Chatters and C R Hajarnavis
45. Moduli, deformations and classifications of compact complex manifolds
    D Sundararaman
46. Nonlinear problems of analysis in geometry and mechanics
    M Atteia, D Bancel and I Gumowski
47. Algorithmic methods in optimal control
    W A Gruver and E Sachs
48. Abstract Cauchy problems and functional differential equations
    F Kappel and W Schappacher
49. Sequence spaces
    W H Ruckle
50. Recent contributions to nonlinear partial differential equations
    H Berestycki and H Brezis
51. Subnormal operators
    J B Conway
52. Wave propagation in viscoelastic media
    F Mainardi
53. Nonlinear partial differential equations and their applications: Collège de France Seminar. Volume I
    H Brezis and J L Lions
54. Geometry of Coxeter groups
    H Hiller
55. Cusps of Gauss mappings
    T Banchoff, T Gaffney and C McCrory

56 An approach to algebraic K-theory
    A J Berrick
57 Convex analysis and optimization
    J-P Aubin and R B Vintner
58 Convex analysis with applications in the differentiation of convex functions
    J R Giles
59 Weak and variational methods for moving boundary problems
    C M Elliott and J R Ockendon
60 Nonlinear partial differential equations and their applications: Collège de France Seminar. Volume II
    H Brezis and J L Lions
61 Singular systems of differential equations II
    S L Campbell
62 Rates of convergence in the central limit theorem
    Peter Hall
63 Solution of differential equations by means of one-parameter groups
    J M Hill
64 Hankel operators on Hilbert space
    S C Power
65 Schrödinger-type operators with continuous spectra
    M S P Eastham and H Kalf
66 Recent applications of generalized inverses
    S L Campbell
67 Riesz and Fredholm theory in Banach algebra
    B A Barnes, G J Murphy, M R F Smyth and T T West
68 Evolution equations and their applications
    F Kappel and W Schappacher
69 Generalized solutions of Hamilton-Jacobi equations
    P L Lions
70 Nonlinear partial differential equations and their applications: Collège de France Seminar. Volume III
    H Brezis and J L Lions
71 Spectral theory and wave operators for the Schrödinger equation
    A M Berthier
72 Approximation of Hilbert space operators I
    D A Herrero
73 Vector valued Nevanlinna Theory
    H J W Ziegler
74 Instability, nonexistence and weighted energy methods in fluid dynamics and related theories
    B Straughan
75 Local bifurcation and symmetry
    A Vanderbauwhede
76 Clifford analysis
    F Brackx, R Delanghe and F Sommen
77 Nonlinear equivalence, reduction of PDEs to ODEs and fast convergent numerical methods
    E E Rosinger
78 Free boundary problems, theory and applications. Volume I
    A Fasano and M Primicerio
79 Free boundary problems, theory and applications. Volume II
    A Fasano and M Primicerio
80 Symplectic geometry
    A Crumeyrolle and J Grifone
81 An algorithmic analysis of a communication model with retransmission of flawed messages
    D M Lucantoni
82 Geometric games and their applications
    W H Ruckle
83 Additive groups of rings
    S Feigelstock
84 Nonlinear partial differential equations and their applications: Collège de France Seminar. Volume IV
    H Brezis and J L Lions
85 Multiplicative functionals on topological algebras
    T Husain
86 Hamilton-Jacobi equations in Hilbert spaces
    V Barbu and G Da Prato
87 Harmonic maps with symmetry, harmonic morphisms and deformations of metrics
    P Baird
88 Similarity solutions of nonlinear partial differential equations
    L Dresner
89 Contributions to nonlinear partial differential equations
    C Bardos, A Damlamian, J I Díaz and J Hernández
90 Banach and Hilbert spaces of vector-valued functions
    J Burbea and P Masani
91 Control and observation of neutral systems
    D Salamon
92 Banach bundles, Banach modules and automorphisms of C*-algebras
    M J Dupré and R M Gillette
93 Nonlinear partial differential equations and their applications: Collège de France Seminar. Volume V
    H Brezis and J L Lions

Similarity solutions of nonlinear partial differential equations

Lawrence Dresner

Oak Ridge National Laboratory, Oak Ridge, Tennessee

Similarity solutions of nonlinear partial differential equations

Pitman Advanced Publishing Program
BOSTON · LONDON · MELBOURNE

PITMAN BOOKS LIMITED
128 Long Acre, London WC2E 9AN

PITMAN PUBLISHING INC
1020 Plain Street, Marshfield, Massachusetts 02050

Associated Companies
Pitman Publishing Pty Ltd, Melbourne
Pitman Publishing New Zealand Ltd, Wellington
Copp Clark Pitman, Toronto

© Oak Ridge National Laboratory 1983

First published 1983

AMS Subject Classifications: (main) 35-02, 35A25, 35G25
　　　　　　　　　　　　　(subsidiary) 35G30, 35B99, 35C99

Library of Congress Cataloging in Publication Data

Dresner, Lawrence
　Similarity solutions of nonlinear partial differential
equations.

　　(Research notes in mathematics; 88)
　　"Pitman advanced publishing program."
　　Bibliography: p.
　　Includes index.
　　1. Differential equations, Partial—Numerical
solutions.　2. Initial value problems—Numerical
solutions.　I. Title.　II. Series.
QA377.D765　1983　　　515.3′53　　　83-7404
ISBN 0-273-08621-9

British Library Cataloguing in Publication Data

Dresner, Lawrence
　Similarity solutions of nonlinear partial differential
　equations.—(Research notes in mathematics; 88)
　1. Differential equations, Partial
　2. Differential equations, Nonlinear
　I. Title　II. Series
　515.3′53　　QA373

　ISBN 0-273-08621-9

All rights reserved. No part of this publication may be reproduced,
stored in a retrieval system, or transmitted, in any form or by any
means, electronic, mechanical, photocopying, recording and/or
otherwise, without the prior written permission of the publishers.
This book may not be lent, resold, hired out or otherwise disposed
of by way of trade in any form of binding or cover other than that
in which it is published, without the prior consent of the publishers.

Reproduced and printed by photolithography
in Great Britain by Biddles Ltd, Guildford

To
Blanche,
Steven, Faye, David, and Eva.

"My simple art ... is but systematized common sense."

                                            A. Conan Doyle, "The Adventure of the Blanched Soldier"

Contents

PREFACE

1 INTRODUCTION 1

2 ORDINARY DIFFERENTIAL EQUATIONS 5
   2.1 First-order ordinary differential equations 5
   2.2 Lie's formula for the integrating factor 6
   2.3 Example 7
   2.4 Analysis of the direction field 8
   2.5 Separatrices 10
   2.6 Separation of variables 11
   2.7 Second-order ordinary differential equations 12
   2.8 Example: Emden-Fowler equations 14

3 LINEAR DIFFUSION 17
   3.1 Birkhoff's idea 17
   3.2 Recapitulation 21
   3.3 Concentration polarization 21

4 NONLINEAR DIFFUSION 25
   4.1 Boltzmann's problem: $c_t = (cc_z)_z$ 25
   4.2 Clamped flux 27
   4.3 The associated group 31
   4.4 Clamped concentration (temperature) 33
   4.5 $c_t = (c^n c_z)_z$ 34
   4.6 Exceptional solutions 35
   4.7 $cc_t = c_{zz}$ 35
   4.8 $c^n c_t = c_{zz}$ 41
   4.9 Transient heat transfer in superfluid helium 41
   4.10 Clamped flux 42
   4.11 Clamped temperature 45
   4.12 Instantaneous heat pulse 47
   4.13 Isothermal percolation of a turbulent liquid into a porous half-space 47

	4.14	Other groups	50
	4.15	Resemblance to dimensional analysis	52
5	BOUNDARY-LAYER PROBLEMS		55
	5.1	Prandtl-Blasius problem of a flat plate	55
	5.2	Blasius's differential equation	57
	5.3	The associated differential equation	59
	5.4	Flat plate with uniform suction or injection	63
	5.5	Thermal boundary layers	65
	5.6	Free convection boundary layer	67
6	WAVE PROPAGATION PROBLEMS		69
	6.1	Introduction	69
	6.2	von Karman-Duwez-Taylor problems	70
	6.3	Elastic (Hookean) wire	71
	6.4	Non-Hookean wire	74
	6.5	Characteristics and Riemann invariants	77
	6.6	Shock conditions	81
	6.7	"Superelastic" wire	82
	6.8	Transverse waves	87
	6.9	Elastic (Hookean) wire - transverse waves	92
	6.10	Long waves in a channel	95
	6.11	Travelling waves	97
7	MISCELLANEOUS TOPICS		101
	7.1	Approximate solutions: diffusion in cylindrical geometry	101
	7.2	Diffusion in cylindrical geometry (cont'd): clamped temperature	105
	7.3	The method of local similarity	107
	7.4	Eigenvalue problems	111
	PROBLEMS		113
	SOLUTIONS TO PROBLEMS		116
	REFERENCES		121
	INDEX		123

Preface

The method of similarity solutions for solving nonlinear partial differential equations has produced a plentiful harvest of results since Birkhoff called attention to it in 1950. The general references in the bibliography at the end of this book cite hundreds of successful individual works. In spite of this success, the method is still not as widely known as equally fruitful methods for solving linear partial differential equations, e.g., separation of variables or Laplace transforms. To popularize the method of similarity solutions, to teach it as a practical technique, to make it a part of the daily armament of the technologist, is the goal of this book.

In harmony with this goal, I have kept the book short and refrained from any great mathematical rigor. To keep the pace swift, I have avoided making an exhaustive survey of the literature. The illustrative examples are drawn mainly from my own work. For most partial differential equations, I have sketched as briefly as possible the physical background. I have not dwelt on it overmuch because our concern here is with the mathematics, but I have not omitted it either because I believe that knowing the physical background helps make clear why we proceed as we do.

This book should be of interest to scientific researchers and practitioners, graduate students, and senior honors undergraduates in the natural sciences, engineering and mathematics. It can serve as the basis of a 30-hour course of lectures at the graduate or senior honors level and has already done so in the Professional Education Program of the Oak Ridge National Laboratory.

To read this book, all one needs to know is the calculus and something about differential equations. Anyone with a good undergraduate education in science, engineering or mathematics should have no trouble, though it will help if he is strong in calculation.

I wish to express my gratitude to Drs Y. Obata and S. Shimamoto of the Japan Atomic Energy Research Institute for their hospitality during my sojourn at their laboratory in 1981-82. It was the comfort and tranquility of my surroundings there that enabled me to do most of the thinking that underlies this book. To Mr M. S. Lubell of the Magnetics and Superconductivity Section of Oak Ridge National Laboratory go my thanks for his encouragement and appreciation and for making available to me the time and resources to prepare the manuscript. And finally my thanks go to the Reports Office of the Fusion Energy Division of Oak Ridge National Laboratory for their meticulous editorial work.

                                                           Lawrence Dresner

Oak Ridge, Tennessee
January 1983

1 Introduction

Everyone can remember from his school days that proving geometrical theorems about triangles or dynamical theorems about planetary orbits is easier if the triangles are equilateral or the orbits circular. Symmetrical problems are the simplest ones to solve. The symmetry of the equilateral triangle and that of the circle are obvious because these figures can readily be visualized. Less obvious but equally real is the algebraic symmetry of ordinary and partial differential equations, which, if present, can facilitate the solution of the equations just as geometrical symmetry can facilitate the solution of geometrical problems. How to exploit the algebraic symmetry of differential equations is what this book is about.

What exactly does 'algebraic symmetry' mean? According to usage long established in mathematics, an object has symmetry if performing certain operations on it leaves it looking the same. For example, rotating an equilateral triangle by 120° around its centroid does not change its appearance. If we hide our eyes, we cannot tell afterwards if the triangle has been rotated. We express the symmetry of the triangle by saying it is *invariant* to rotation of 120° around its centroid. It is invariant to other transformations as well: if we reflect it about one of its altitudes, the image looks the same as the object. The circle, even more symmetric than the triangle, is invariant to all rotations around its centroid and to reflection in any diameter.

Differential equations, both ordinary and partial, are sometimes invariant to groups of algebraic transformations, and these algebraic invariances, like the geometric ones mentioned above, are also symmetries. About a hundred years ago, the Norwegian mathematician Sophus Lie hit upon the idea of using the algebraic symmetry of ordinary differential equations to aid him in their solution [CO31, DI23]. In the course of his work, he achieved two profoundly important results: he showed how to use knowledge of the transformation group (1) to construct an integrating factor for first-order ordinary differential equations and (2) to reduce second-order ordinary differential equations to first order by a change of variables. These two results are all the more important because *they do not depend on the equation's being linear*. Both play central roles in this book.

Also about a hundred years ago, the Austrian physicist Boltzmann [BO94] used the algebraic symmetry of the partial differential diffusion equation to study diffusion with a concentration-dependent diffusion coefficient. To my knowledge, Boltzmann made no explicit mention of transformation groups or symmetry, but his procedure was the same one that is used today. The crux of his method is using the symmetry to find special solutions of the partial differential equation by solving a related *ordinary* differential equation.

The American mathematician Garrett Birkhoff was first to recognize that Boltzmann's procedure depended on the algebraic symmetry of the diffusion equation and could be generalized to other partial differential equations, *including nonlinear ones* [BI50]. Using the algebraic symmetry of the partial differential equation, he showed how solutions can be found merely by solving a related ordinary differential equation, a much easier task. For reasons that become clear later, such solutions are called similarity solutions.

After Birkhoff's work called attention to them, similarity solutions were found for a great many physical problems in such diverse fields as heat and mass transfer, fluid dynamics, solid mechanics, applied superconductivity and plasma physics. In most of these problems two common features recurred. First, the transformation groups leaving the partial differential equation invariant were *families* of *stretching* groups G of the form $x' = \lambda^\alpha x$, $y' = \lambda^\beta y,\ldots$, $0 < \lambda < \infty$. Second, the partial differential equations were of second order.

In problems with the second feature, the related ordinary differential equation is also of second order. In some cases, this second-order ordinary differential equation is integrable in terms of elementary or tabulated functions, but in most cases it is not. Some time ago, I observed that in problems with the first feature also, the second-order ordinary differential equation is invariant to a stretching group G' related to the family of groups G to which the partial differential equation is invariant [DR80]. By using Lie's second theorem mentioned above, then, the second-order ordinary differential equation can be reduced to first order by a change of variables. The first-order equation can be analyzed very conveniently by studying its direction field.

At this point, it will help to introduce some abbreviations and new terms.

Henceforth, the abbreviations pde, ode and deq will stand for partial differential equation, ordinary differential equation and differential equation, respectively. The second-order pde under consideration, being the only one involved in the problem, will be called simply the pde. The family of stretching groups G to which the pde is invariant will be called the principal group, and the second-order ode obtained with the help of the principal group will be called the principal deq. The stretching group G' to which the principal ode is invariant will be called the associated group, and the first-order ode obtained with the help of the associated group will be called the associated deq. Figure 1 summarizes this scheme of terminology.

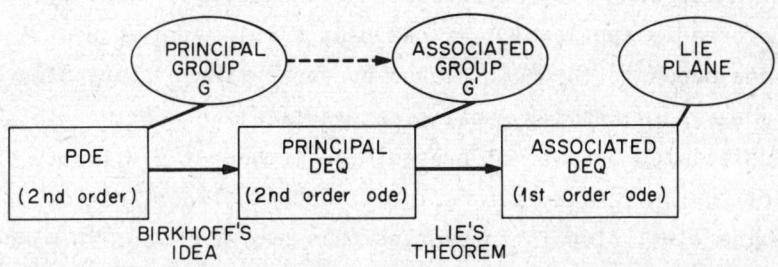

Fig. 1. Scheme of terminology used in this book

The direction field of the associated deq bears a relation to the principal deq similar to that which the phase plane bears to an autonomous second-order deq of motion. For this reason, and because the associated deq is obtained with the help of Lie's theorem, about ten years ago I proposed calling this direction field the Lie plane [DR71]. This name is used in this book.

Not every physical problem describable by the pde is solvable by the procedure sketched in Fig. 1, i.e., not every solution of the pde is a similarity solution. Whether a particular problem has a similarity solution depends on the boundary and initial conditions. All other writers on this subject have concentrated on finding physical problems having similarity solutions and have usually solved the principal deq numerically. In other words, they have concentrated exclusively on the left half of Fig. 1. In this book I place emphasis on the right half of Fig. 1 and especially on analysis of the Lie plane.

Several substantial benefits arise from the study of the Lie plane.

First, the singular points of the associated ode and the separatrices that may join them occasionally provide the limiting behaviours of the similarity solution. The illustrative problems in this book contain several practical results obtained in this way. Second, when the solution of the principal deq is determined by two-point boundary conditions, study of the Lie plane sometimes enables us to avoid the trial and error of the usual shooting method. It also allows us to decide the stable direction of numerical integration when that is necessary. And last, study of the Lie plane sometimes allows us to identify special solutions of the pde that might otherwise go unnoticed.

Because of these benefits and because the method of similarity solutions is simple and broadly applicable, it can play a role with respect to nonlinear pdes analogous to the role played by separation of variables with respect to linear pdes, namely, that of a practical workhorse. This is not now widely appreciated because at present the method of similarity solutions is not part of our university curricula. I hope that publication of this book will be the first step in correcting this oversight and in placing in the practitioner's hands an important tool.

2 Ordinary differential equations

2.1 First-order odes

First-order odes differ from higher-order odes in that the condition they express can be visualized directly: the deq $f(x,y,\dot{y}) = 0$ expresses a relation between the derivative $\dot{y} = dy/dx$ and the coordinates x, y that we can plot by drawing a short line segment with the slope $\dot{y}$ at each point x, y. From such a plot we can see at a glance what kind of solutions the deq has. Such a plot is called the direction field of the deq.

Figure 2 shows by way of illustration the direction field of the simple equation $\dot{y} = x$. The solutions $y(x)$ of the deq are curves in the (x,y)-plane tangent at every point to the local line segment (integral curves). It is easy to see that in Fig. 2 these integral curves are a one-parameter family. The parameter labeling the individual curves can be their y-intercept, for example.

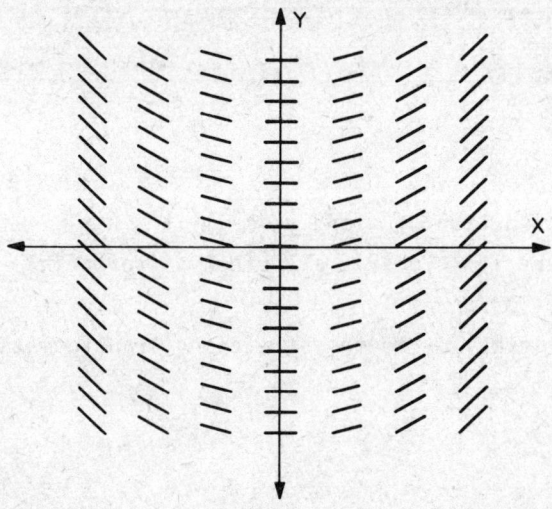

Fig. 2. Direction field of the deq $\dot{y} = x$

The deq $\dot{y} = x$ is invariant to the group of transformations $y' = \lambda^2 y$, $x' = \lambda x$, where λ can have any value between 0 and ∞. This means if we substitute for x and y their values in terms of x' and y', the resulting deq in x' and y' is the same as the original deq in x and y. The integral curve $y = x^2/2 + c$, labeled by the y-intercept c, on the other hand, is transformed into the integral curve $y' = x'^2/2 + \lambda^2 c$, labeled by another y-intercept $c' = \lambda^2 c$. But a moment's thought shows that these transformations of the integral curves among themselves leave the *family* of integral curves invariant; for if we plot the integral curves on one piece of paper and their images on another piece of paper, when we are done the two plots look the same. The deq is the family, so it must remain invariant, even though individual integral curves do not.

2.2 Lie's formula for the integrating factor

One conventional representation of the deq $f(x,y,\dot{y}) = 0$ is

$$M(x,y)dx + N(x,y)dy = 0. \tag{1}$$

If the one-parameter family of integral curves of (1) is represented by $\phi(x,y) = c$, then along any such curve

$$\phi_x \, dx + \phi_y \, dy = 0 \quad \left(\phi_x \equiv \frac{\partial \phi}{\partial x}, \; \phi_y \equiv \frac{\partial \phi}{\partial y}\right). \tag{2}$$

From (1) and (2) we see that

$$\frac{\phi_x}{M} = \frac{\phi_y}{N}. \tag{3}$$

The equal ratios (3) are a function of x and y; call it $\mu(x,y)$. If we multiply (1) by $\mu(x,y)$, we convert it into (2), a perfect differential. So μ is an integrating factor for (1).

Suppose now that (1) is invariant to the stretching transformations

$$\left.\begin{array}{l} y' = \lambda^\beta y \\ x' = \lambda x \end{array}\right\} \quad 0 < \lambda < \infty \tag{4}$$

(Note that we lose no generality by choosing the exponent of λ in the second equation to be 1.) Then individual integral curves transform into other integral curves:

$$\phi(\lambda x, \lambda^\beta y) = c(\lambda) \tag{5}$$

If we differentiate (5) with respect to λ and set $\lambda = 1$, we get

$$x\phi_x + \beta y \phi_y = \left(\frac{dc}{d\lambda}\right)_{\lambda=1} \tag{6}$$

which can be written as

$$x(\mu M) + \beta y(\mu N) = \left(\frac{dc}{d\lambda}\right)_{\lambda=1}. \tag{7}$$

Thus

$$\mu = \frac{(dc/d\lambda)_{\lambda=1}}{xM + \beta yN}. \tag{8}$$

Since $(dc/d\lambda)_{\lambda=1}$ is a constant, we can ignore it and use the expression $(xM + \beta yN)^{-1}$ as an integrating factor.

Lie did not restrict himself to stretching transformations but dealt with more general groups of transformations. He obtained a formula, of which (8) is a special case, applicable to any group. Since the groups dealt with in this book are stretching groups, we have no need for the more general formula.

2.3 Example

As an example, let us consider the deq

$$\dot{y} = \frac{y(x - y^2)}{x^2} \tag{9}$$

which can be written in the standard form (1) as follows

$$y(y^2 - x)dx + x^2 dy = 0. \tag{10}$$

It is invariant to the group of transformations

$$\left.\begin{array}{l} y' = \lambda^{1/2} y \\ x' = \lambda x \end{array}\right\} \quad 0 < \lambda < \infty \tag{11}$$

According to Lie's theorem

$$\mu = \left(xy(y^2 - x) + \frac{1}{2}x^2 y\right)^{-1} = \left(xy^3 - \frac{1}{2}x^2 y\right)^{-1} \tag{12}$$

is an integrating factor. Then

$$\frac{\partial \phi}{\partial y} = \frac{x^2}{xy^3 - \frac{1}{2}x^2 y} = -\frac{2}{y} + \frac{2y}{y^2 - \frac{1}{2}x} \tag{13}$$

so that

$$\phi = -2\ln y + \ln\left(y^2 - \frac{1}{2}x\right) + F(x) \tag{14}$$

where $F(x)$, an arbitrary function of x, plays the role of integration constant. Differentiating partially with respect to x we find

$$\frac{\partial \phi}{\partial x} = \frac{dF}{dx} - \frac{1}{2y^2 - x} . \tag{15}$$

Comparing this with μM, we see that

$$\frac{dF}{dx} = \frac{1}{x} \tag{16}$$

so that

$$F(x) = \ln x + \ln A \tag{17}$$

where $\ln A$ is an integration constant. Then

$$\phi(x,y) = \ln \frac{A\, x\left(y^2 - \frac{1}{2}x\right)}{y^2} . \tag{18}$$

If we set $\phi(x,y)$ equal to a constant, and write for convenience $e^\phi/A = -c/2$, where c is a new constant, we can solve (18) for y and find

$$y = \frac{x}{\sqrt{2x + c}} . \tag{19}$$

The reader may verify by differentiation that (19) satisfies (9). In so doing, let the reader note that there are no restrictions on c so that it may be positive or negative.

2.4 Analysis of the direction field

With the formula (19) available to help us check our conclusions, let us see what we can learn about the solutions of (9) from its direction field, proceeding as though we did not have a closed formula for the integral curves.

In analyzing a direction field, the first thing to do is to locate the curves on which dy/dx = 0 or ∞, i.e., on which M = 0 or N = 0. Only on these lines does dy/dx change sign so, in the regions between, the sign of dy/dx is fixed. Shown in Fig. 3 is a sketch of the upper half of the direction field

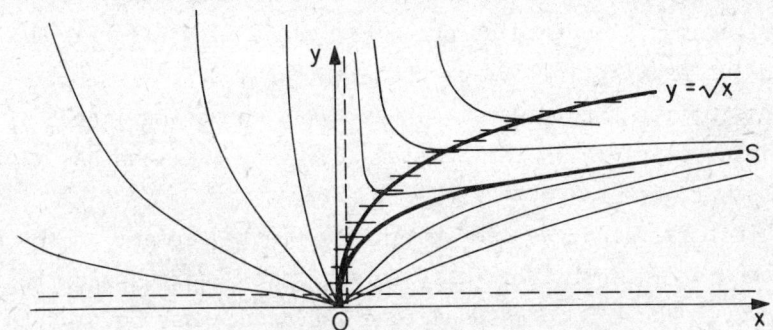

Fig. 3. Sketch of the direction field and integral curves (light lines) of Eq. (9)

and integral curves of Eq. (9). Since (9) is invariant to the transformation y' = -y, x' = x, the upper and lower halves of the direction field are mirror images of one another; thus we need consider only the upper half. The slope $\dot{y}$ is zero on the x-axis and on the curve $y = \sqrt{x}$; it is infinite on the y-axis. It is easy to see from (9) that the slope has the sign in each region shown in the diagram.

Only one integral curve passes through each point in the direction field, with the possible exception of those points at which both M and N simultaneously vanish. Such points, at which many slopes are possible, are called singular points. The origin O in Fig. 3 is such a singular point. Do integral curves pass through O? If so, how do they behave?

Near the origin both x and y become very small. Along any integral curve approaching the origin, one of three mutually exclusive alternatives must hold, namely, $y \ll \sqrt{x}$, $y \sim \sqrt{x}$, and $y \gg \sqrt{x}$. If $y \gg \sqrt{x}$, then (9) becomes $\dot{y} = -y^3/x^2$, which is easily solvable. It gives $1/2y^2 + 1/x = A$, a constant. Thus when x becomes <1/A, y can no longer be real. This contradiction rules out the possibility of integral curves on which $y \gg \sqrt{x}$ approaching the origin.

If $y \ll \sqrt{x}$, the deq (9) becomes $\dot{y} = y/x$, which again is easily solvable and gives y = Bx, B = a constant. This behaviour is consistent with the hypothesis $y \ll \sqrt{x}$ and is therefore possible. If $y \sim \sqrt{x}$, we can set $y = D\sqrt{x}$ and substitute in (9). We find that D can only have the value $\sqrt{2}/2$, so we have found the special solution to (9), $y = \sqrt{x/2}$. It is shown in Fig. 3 marked S.

The same three possibilities exist as $x \to \infty$. But now only the possibility $y = \sqrt{x/2}$ avoids a contradiction. This means all integral curves in the first quadrant asymptotically approach the curve S.

The only possibility for integral curves lying above the line S is $y \gg \sqrt{x}$, so that each integral curve in this region crosses $y = \sqrt{x}$ and has a vertical asymptote at some finite value of x.

With these facts at hand, we can sketch the first quadrant of the direction field shown in Fig. 3. It is easy to see that S corresponds to $c = 0$, the integral curves above S correspond to $c < 0$, and those below S to $c > 0$. A similar analysis shows that the second quadrant must also look as sketched, and reference to (19) shows that in the second quadrant $c > 0$ and the negative sign of the square root must be used.

2.5 Separatrices

According to transformation (11), $\dot{y}' = \lambda^{-1/2} \dot{y}$. Therefore, integral curves whose slope is positive everywhere can only transform into other integral curves whose slope is positive everywhere. The same is true for integral curves whose slope is negative everywhere, and also for integral curves whose slope is zero somewhere. This means that the integral curves in the first quadrant lying below S transform into themselves, whose lying above S transform into themselves, and those lying in the second quadrant transform into themselves. The separatrix S, being the only integral curve having an infinite slope at the origin, must be its own image under (11). This is easily verified.

Being its own image is a general property of separatrices that we can employ to calculate them without solving the differential equation. If $g(x,y) = 0$ represents the separatrix, then $g(x',y') = 0$ does as well. Thus on the separatrix

$$g(\lambda x, \lambda^\beta y) = 0. \tag{20}$$

Again we differentiate with respect to λ and set $\lambda = 1$. We can rearrange the result in the form

$$-\frac{\partial g/\partial x}{\partial g/\partial y} = \frac{\beta y}{x} \tag{21}$$

Now the left-hand side of (21) is the slope dy/dx of $g(x,y)$ at the point x,y. The separatrix, lying infinitesimally close to other integral curves, must also be an integral curve. Hence, it must be given by the equivalent *algebraic* equations

$$f\left(x, y, \frac{\beta y}{x}\right) = 0 \tag{22}$$

or

$$xM + \beta yN = 0. \tag{23}$$

In the case of deq (10), (23) gives $x = 0$, $y = 0$, and $y = \sqrt{x/2}$ as invariant curves, i.e., curves that are their own images. All are separatrices.

2.6 Separation of variables

An alternative to introducing an integrating factor in first-order odes is to separate variables. Here, too, group invariance can help. In general, to separate variables one must introduce new "canonical" coordinates calculable from the transformation group of the ode. But in the case of a stretching group, it is sufficient to replace the dependent variable by a *group invariant* $u = y/x^\beta$. For, suppose we write the deq in the form

$$\dot{y} = F(x,y) \tag{24}$$

Then,

$$\frac{du}{dx} = \frac{F}{x^\beta} - \beta \frac{y}{x^{\beta+1}} = \frac{1}{x}\left(\frac{F}{x^{\beta-1}} - \beta u\right). \tag{25}$$

Now when x and y transform according to (4), then $\dot{y}' = \lambda^{\beta-1}\dot{y}$. The invariance of (24) to the transformation (4) means

$$\lambda^{\beta-1}\dot{y} = F(\lambda x, \lambda^\beta y) \tag{26}$$

Differentiating (26) with respect to λ and setting $\lambda = 1$, we find

$$xF_x + \beta y F_y = (\beta - 1)F. \tag{27}$$

The standard method of finding the general solution of (27) is to find two integrals of the associated characteristic equations

$$\frac{dx}{x} = \frac{dy}{\beta y} = \frac{dF}{(\beta - 1)F} \tag{28}$$

and set one equal to an arbitrary function of the other. Two such integrals are $F/x^{\beta-1}$ and y/x^β. Thus, most generally,

$$\frac{F}{x^{\beta-1}} = G\left(\frac{y}{x^\beta}\right) = G(u). \tag{29}$$

Then we can write (25) as

$$\frac{dx}{x} = \frac{du}{G(u) - \beta u} \tag{30}$$

in which the variables are separated.

It follows from (30) that the variables will also separate if we use any arbitrary function of u as the new dependent variable. If the reader will return now to the example of Eq. (9) and use the invariant y^2/x as the new dependent variable, he will find the labour of calculation required to obtain (19) somewhat reduced.

2.7 Second-order odes

The idea of using a group invariant as a new variable can be extended to second-order odes. Lie proposed using an *invariant* $u(x,y)$ and a *first differential invariant* $v(x,y,\dot{y})$ as new independent and dependent variables. For the stretching group (4), $u = y/x^\beta$ and $v = \dot{y}/x^{\beta-1}$ are an invariant and a first differential invariant. Differentiating them along an integral curve we find

$$x \frac{dv}{dx} = \frac{\ddot{y}}{x^{\beta-2}} - (\beta - 1)\frac{\dot{y}}{x^{\beta-1}} = \frac{\ddot{y}}{x^{\beta-2}} - (\beta - 1)v \tag{31a}$$

$$x \frac{du}{dx} = \frac{\dot{y}}{x^{\beta-1}} - \beta \frac{y}{x^\beta} = v - \beta u \tag{31b}$$

so that

$$\frac{dv}{du} = \frac{\ddot{y}/x^{\beta-2} - (\beta - 1)v}{v - \beta u} \tag{31c}$$

Suppose we write the second-order ode as

$$\ddot{y} = F(x,y,\dot{y}) \tag{32}$$

Invariance of this deq to the transformations (4) means

$$\lambda^{\beta-2}\ddot{y} = F(\lambda x, \lambda^\beta y, \lambda^{\beta-1}\dot{y}). \tag{33}$$

Again differentiating with respect to λ and setting $\lambda = 1$, we find

$$xF_x + \beta y F_y + (\beta - 1)\dot{y} F_{\dot{y}} = (\beta - 2)F \tag{34}$$

The characteristic equations of (34) are

$$\frac{dx}{x} = \frac{dy}{\beta y} = \frac{d\dot{y}}{(\beta - 1)\dot{y}} = \frac{dF}{(\beta - 2)F} \tag{35}$$

Equations (35) have three independent integrals which can be taken as $y/x^\beta = u$, $\dot{y}/x^{\beta-1} = v$, and $F/x^{\beta-2}$. The most general solution of (34) is obtained by setting one of these integrals equal to an arbitrary function of the other two, namely,

$$F/x^{\beta-2} = G(u,v). \tag{36}$$

Noting that $\ddot{y} = F$, we see that in view of (36), we can write (31c) as

$$\frac{dv}{du} = \frac{G(u,v) - (\beta - 1)v}{v - \beta u} \tag{37}$$

a *first-order* deq for v in terms of u.

If we succeed in integrating (37) we then obtain v as a function of u. This relation represents a first-order deq for $\dot{y}$ in terms of x and so must be integrated again to obtain y as a function of x. But it, too, is invariant to (4) and so may be dealt with either by applying an integrating factor or by separating variables. If we cannot integrate the first-order deq (37), by far the commonest case, we proceed by studying its direction field. In this way, we can learn much useful information about the solutions of the second-order deq (32).

2.8 Example: Emden-Fowler equation

The Emden-Fowler equation

$$\ddot{y} + \frac{2\dot{y}}{x} + y^n = 0 \tag{38}$$

arises in the study of the equilibrium mass distribution in a cloud of gas with adiabatic exponent $(n+1)/n$ held together by gravitational force. Equation (38) is invariant to a stretching group like (4) with $\beta = -2/(n-1)$. Taking $u = yx^{2/(n-1)}$ and $v = \dot{y}x^{(n+1)/(n-1)}$, we find

$$\frac{dv}{du} = - \frac{(n-1)u^n + (n-3)v}{(n-1)v + 2u}. \tag{39}$$

When $n = 5$, (39) is integrable and gives

$$3uv + 3v^2 + u^6 = \text{constant} \tag{40a}$$

or, in terms of x and y,

$$3x^2 y\dot{y} + 3x^3 \dot{y}^2 + x^3 y^6 = \text{constant}. \tag{40b}$$

Now the solutions we want must be finite and have zero derivative at the origin $x = 0$ because y is the gravitational potential. So the constant in (40b) must be zero.

Equation (40b) is also invariant to group (4) with $\beta = -1/2$ ($n = 5$). If we introduce the invariant $w = u^2 = xy^2$, we can separate variables. After a short calculation, we find

$$\frac{dx}{x} = \frac{\sqrt{3}}{2} \frac{dw}{w\left(\frac{3}{4} - w^2\right)^{1/2}} \tag{41}$$

which can be integrated after making the substitution $w = (\sqrt{3}/2) \sin(\theta/2)$. After some further computation we find

$$w = \frac{3ax}{x^2 + 3a^2}, \quad a = \text{constant} \tag{42a}$$

or

$$y = \left(\frac{3a}{x^2 + 3a^2}\right)^{1/2}. \tag{42b}$$

This solution is already known [DA 60].

When n = 3, (39) is not simply integrable, and the analysis of its direction field provides an illustration of the subtleties we may encounter in the analysis of nonlinear first-order deqs. Equation (39) becomes

$$\frac{dv}{du} = - \frac{u^3}{u + v} \qquad (43)$$

the direction field of which is shown in Fig. 4.

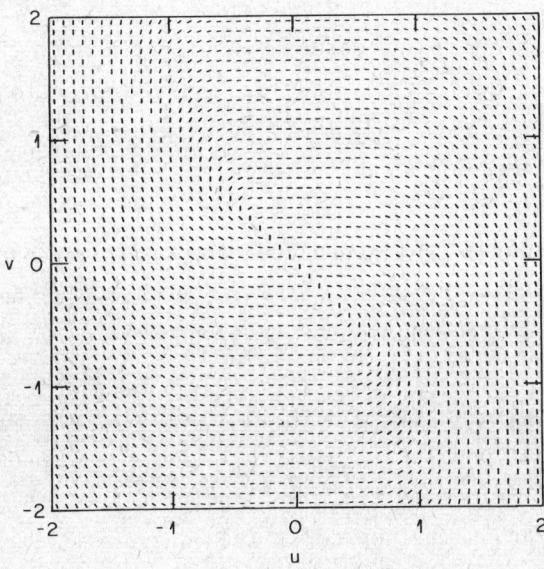

Fig. 4. Direction field of Eq. (43)

The integral curves circle the origin at large enough radii, but once within a critical radius they approach the origin, drawing ever nearer to the line v = -u as they do so. Curves intersecting the line v = -u at abscissas whose absolute value is greater than some value u_0, make one more counterclockwise circuit of the origin, shrinking as they do so; curves intersecting v = -u at abscissas whose absolute value is less than u_0 approach the origin along the line v = -u. What about the exceptional integral curve that intersects the line v = -u at u = $\pm u_0$? It approaches the origin along the u-axis, i.e., with zero slope.

Because y is finite and $\dot{y}$ = 0 at x = 0, the origin u = xy = 0, v = $x^2\dot{y}$ = 0 in the u,v plane must lie on the integral curve of (43) that we want. How do x and y behave as the various integral curves approach the origin? The integral curves that approach the line $v = -u$ near the origin have equations of the form $v = -u + \varepsilon(u)$ where ε has the same sign as u and approaches zero faster than u. Then

$$x \frac{du}{dx} = x \frac{d}{dx}(xy) = x^2\dot{y} + xy = u + v = \varepsilon \tag{44}$$

so that

$$\ln x = \int^u \frac{du}{\varepsilon}. \tag{45}$$

As $u \to 0$, the right-hand side of (45) approaches $-\infty$ so that $x \to 0$. Furthermore, since $\varepsilon \to 0$ faster than u does, x must approach zero faster than u does. But since $y = u/x$, $y \to \pm\infty$ as $x \to 0$, the sign depending on the sign of u. So the integral curves that approach the origin along $v = -u$ cannot be the ones we want.

What about the exceptional curve that approaches the origin with zero slope? Zero slope means that near the origin $u \gg v$, so that (43) becomes $dv/du = -u^2$. Thus $v = -u^3/3$ is the equation of this integral curve near the origin. Written in terms of x and y, this becomes $\dot{y}/x = -y^3/3$ near $x = 0$. This result gives us the otherwise indeterminate limit of $\dot{y}/x$, and substitution into the Emden-Fowler equation itself shows that $\ddot{y}(0) = -y^3(0)/3$.

As we advance away from the origin in the Lie plane, the integral curves spiral out from the origin. Each time they cross the line $u = 0$, $y = 0$; each time they cross the line $v = 0$, $\dot{y} = 0$. So the solutions oscillate as x increases. In the physical application of the Emden-Fowler equation to stellar constitution, only the part out to the first root is of importance.

In this case, we cannot profitably go further without some numerical work, which we shall not undertake here. The points of the present discussion are two: first, to show how the desired solution to a physical problem may correspond to an exceptional solution of the associated first-order ode, and second, how knowing this we may sometimes obtain the limiting behaviour of the desired solution (in this case, $y = y(0) - x^2 y^3(0)/6 + \ldots$). In the solution of problems governed by pdes, such limiting behaviours sometimes provide us with useful physical information.

3 Linear diffusion

3.1 Birkhoff's idea

We saw in Chapter 2 that if a symmetry group of the deq is known, it is possible to calculate an invariant integral curve such as a separatrix algebraically, i.e., without integrating the deq. Of course this procedure is useful only if the invariant curve obeys boundary conditions that make it the solution of a physically interesting problem. Birkhoff [BI50] pointed out that a similar thing occurs for partial differential equations in two independent variables: if we know a symmetry group for the pde, we can calculate invariant solutions by solving a related ode rather than the pde itself. Again this procedure is useful only if the invariant solution describes a physically interesting situation.

Although this idea makes its major impact in the realm of nonlinear pdes, where it has no competitor as widely applicable, it also can be used with great effect on linear pdes. To show the peculiar features of the method, we start with the ordinary diffusion equation

$$c_t = c_{zz} \tag{1}*$$

It is invariant to the group of transformations

$$\begin{aligned} c' &= \lambda^\alpha c \\ t' &= \lambda^2 t \quad 0 < \lambda < \infty \\ z' &= \lambda z \end{aligned} \tag{2}$$

where α can have any value. Any solution of (1), $C = f(z,t)$, will become another solution under the transformation (2). If the image solution is the same as the object, then the solution is invariant to (2). The condition for this is

---

* This is the form the diffusion equation takes in *special units* in which the diffusivity equals 1. We shall make extensive use of such special units in this book. Formulas valid in ordinary units, e.g., mks units, can always be recovered from their more succinct counterparts written in special units.

$$\lambda^\alpha f(z,t) = f(\lambda z, \lambda^2 t) \tag{3}$$

(The way to understand condition (3) is as follows. Suppose the point (z,t,C) lies on an invariant integral surface S, i.e., suppose $C = f(z,t)$. Its image (z', t', C') also lies on S, so that $C' = f(z', t')$. Then from (2) we have $\lambda^\alpha C = f(\lambda z, \lambda^2 t)$. If we now replace C by $f(z,t)$, we get (3).) If we differentiate (3) with respect to λ and set $\lambda = 1$, we get

$$zf_z + 2tf_t = \alpha f. \tag{4}$$

The characteristic equations

$$\frac{dz}{z} = \frac{dt}{2t} = \frac{df}{\alpha f} \tag{5}$$

have the two independent integrals $z/t^{1/2}$ and $f/t^{\alpha/2}$, so the most general form f can take if it is to be invariant to (2) is

$$f = t^{\alpha/2} y\left(\frac{z}{t^{1/2}}\right) \tag{6}$$

where y is an arbitrary function.

If there are invariant solutions they must have the form (6); if we substitute (6) into (1) it will have to satisfy (1). But the partial derivatives of f are expressible in terms of the ordinary derivatives of y, so (1) will become an ode for y. Thus

$$C_t = f_t = t^{(\alpha/2)-1}\left(\frac{\alpha}{2}y - \frac{1}{2}\frac{z}{t^{1/2}}\dot{y}\right) \tag{7a}$$

$$C_z = t^{(\alpha-1)/2}\dot{y} \tag{7b}$$

$$C_{zz} = t^{(\alpha/2)-1}\ddot{y} \tag{7c}$$

and (1) becomes

$$\ddot{y} = \frac{\alpha}{2}y - \frac{1}{2}\frac{z}{t^{1/2}}\dot{y}. \tag{8}$$

Now the quantity $z/t^{1/2}$, which we henceforth abbreviate as x, is the argument of the function y. So (8) can be written as

$$\ddot{y} + \frac{x}{2}\dot{y} - \frac{\alpha}{2}y = 0 \tag{9}$$

an ode for $y(x)$.

In passing from (1) to (9), the various powers of z and t that appeared in the derivatives of C either cancelled or could be combined into powers of x. This could only happen because (6) was written in terms of the correct variables $C/t^{\alpha/2}$ and $z/t^{1/2}$. The central role of the group invariance is to help us discover the correct variables. These correct variables are called similarity variables.

What about the value of α? It is determined by the boundary and initial conditions and will be different for different physical problems. We shall consider several problems here, and we begin with one that is relatively simple, namely, a half-space, in which initially $C = 0$, has its face suddenly raised and held at $C = 1$. (Think of C as temperature.) The boundary and initial conditions for this problem are these:

$$C(0,t) = 1 \quad t > 0 \tag{10a}$$

$$C(z,0) = 0 \quad z > 0 \tag{10b}$$

$$C(\infty,t) = 0 \quad t > 0. \tag{10c}$$

For (6) to satisfy (10a), we must have $\alpha = 0$ and $y(0) = 1$. To satisfy (10b) and (10c), we must have $y(\infty) = 0$. This "collapse" of two conditions for the pde into one condition for the ode is vital. In general, more conditions are needed to specify a solution of a pde than to specify a solution of an ode. If some of the conditions for the pde do not collapse to the same condition for the ode, there may be too many mutually inconsistent conditions to determine a solution of the ode. What this means simply is that the pde has more solutions than those that can be represented in the form (6).

For the clamped-temperature problem we now have

$$\ddot{y} + \frac{x}{2}\dot{y} = 0 \tag{11a}$$

$$y(0) = 1, \quad y(\infty) = 0. \tag{11b}$$

This equation can easily be integrated twice to give

$$y = \mathrm{erfc}\left(\frac{x}{2}\right) \tag{12}$$

where erfc is the complementary error function. It follows from (12), furthermore, that $C_z(0,t) = -(\pi t)^{-1/2}$ so the heat flux through the front surface falls as $t^{-1/2}$. This result, without the value of the constant $\pi^{-1/2}$, could have been deduced from (7b) without solving the ode (11a).

When $\alpha = 1$, we have the case in which the heat flux at the wall is suddenly clamped, for then $C_z(0,t) = \dot{y}(0)$, a constant independent of t. Then (9) becomes

$$\ddot{y} + \frac{x}{2}\dot{y} - \frac{1}{2}y = 0. \tag{13}$$

Now $y = x$ is a special solution of (13), and because (13) is linear, we can find a second, independent solution by the classical procedure of setting $y = wx$. Then we find the following *separable* equation for w:

$$\frac{\ddot{w}}{\dot{w}} = -\frac{2}{x} - \frac{x}{2}. \tag{14}$$

Then

$$y = wx = Ax \int_x^\infty e^{-x^2/4} \frac{dx}{x^2} = A\left(e^{-x^2/4} - \frac{x}{2}\int_x^\infty e^{-x^2/4} dx\right) \tag{15}$$

where A is a constant of integration. If we choose $\dot{y}(0) = -1$, then we must take $A = 2/\sqrt{\pi}$.

When $\alpha \neq 0,1$ (9) can be solved in terms of the confluent hypergeometric function:

$$y = \exp\left(-\frac{x^2}{4}\right) U\left(\frac{\alpha + 1}{2}, \frac{1}{2}, \frac{x^2}{4}\right) \tag{16}$$

where U is the function defined by M. Abramowitz and I. A. Stegun [AB68, p. 504, 13.1.3]. The reader can verify this by substitution. It follows after a straightforward calculation that

$$-\frac{\dot{y}(0)}{y(0)} = \frac{\Gamma\left(\frac{\alpha + 2}{2}\right)}{\Gamma\left(\frac{\alpha + 1}{2}\right)}, \tag{17}$$

which agrees with the results $1/\sqrt{\pi}$ and $\sqrt{\pi}/2$ we just found in the cases $\alpha = 0$ and $\alpha = 1$, respectively. Because the diffusion equation is linear and solutions can be superposed, we can use (17) to find the heat flux into the surface $z = 0$ when the time dependence of the temperature there is expressed as an arbitrary sum of powers of t. Conversely, given the heat flux as such a sum, we can find the temperature. Superposability is the most important thing we lose when we deal with nonlinear equations.

The case $\alpha = -1$ corresponds to heating of the whole space by an instantaneous unit heat pulse delivered at the plane $z = 0$ at time $t = 0$, for conservation of energy requires

$$1 = \int_{-\infty}^{+\infty} C\, dx = t^{\alpha/2} \int_{-\infty}^{+\infty} y\left(\frac{z}{t^{1/2}}\right) dz \qquad (18)$$

and the right-hand side can only be independent of t if $\alpha = -1$, in which case it becomes $\int_{-\infty}^{+\infty} y(x)dx$. Equation (9) can easily be integrated when $\alpha = -1$, and gives $y = $ constant $\times \exp(-x^2/4)$. To satisfy the requirement of unit integral over all space, we must choose the constant to be $(4\pi)^{-1/2}$, so that

$$C = \frac{\exp(-x^2/4)}{(4\pi t)^{1/2}} \qquad (19)$$

a well-known result.

3.2 Recapitulation

The essential points of the foregoing procedure are the following:

(1) Substitution of a trial solution such as (6) made up of group invariants reduces the pde to an ode. Such solutions are called similarity solutions because they have the property that the profiles of C versus z at various t are geometrically similar, i.e., can be obtained from one another by stretching ordinate and abscissa appropriately.

(2) The values of any remaining constants such as α are determined by the boundary conditions (all of which, it goes without saying, must also be invariant to the transformation group).

(3) Some of the boundary conditions "collapse" to the same condition for the ode so that its solution is not overdetermined.

3.3 Concentration polarization

A slightly more complex but still linear diffusion problem arises in the desalting of water by reverse osmosis, i.e., by its passage under pressure through a semi-permeable membrane [SH65, DR64]. The retained salt builds up against the membrane, a phenomenon called concentration polarization. The excess salt concentration at the wall is a valuable datum in process design. Figure 5 shows a sketch of a slit-shaped channel. The salty feed water

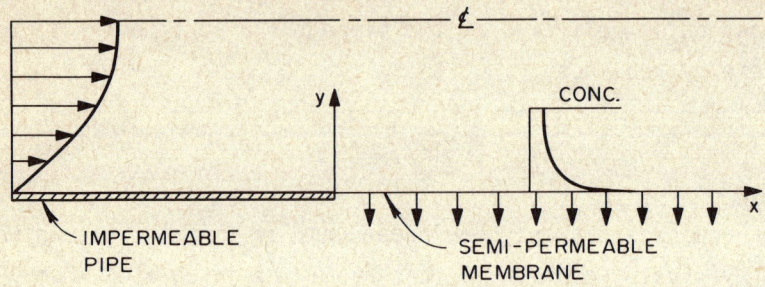

Fig. 5. Sketch explaining concentration polarization in steady-state reverse osmosis: entrance region

flowing in from the left is assumed to be in fully developed laminar flow.

In the entrance region, the excess salt concentration near the membrane is governed by the partial differential equation

$$C_{yy} = yC_x \qquad (20)^*$$

and the boundary conditions

$$(C_y)_{y=0} = -1 \quad x > 0 \qquad (21a)$$

$$C(0,y) = 0 \quad y > 0 \qquad (21b)$$

$$C(x,\infty) = 0 \quad x > 0 \qquad (21c)$$

Setting $y = \infty$ in (21c) is allowed because in the entrance region the thickness of the concentration boundary layer is very much smaller than the half-thickness of the channel. In a manner of speaking, then, the salt does not "know" that the channel has an opposite wall. This kind of approximation can often be made in limiting cases so that the limiting behaviours of the solution can be expressed as similarity solutions. This point has been especially

---

* In special units in which x is measured in units of $u_y D^2/v^3$, y is measured in units of D/v, and C is measured in units of C_0, where D is the diffusivity of the salt, v is the permeation velocity of the water through the membrane, u_y is the axial flow velocity gradient at the wall, and C_0 is the salt concentration of the feedwater.

stressed by Barenblatt [BA79].

These equations and boundary conditions are invariant to the group of transformations

$$C' = \lambda^\alpha C$$
$$x' = \lambda^3 x \qquad (22)$$
$$y' = \lambda y.$$

(We can see now that if a finite value of y appeared in (21c) instead of infinity, (21c) would not be invariant to (22). So the results we derive here are only applicable to the entrance region, where the boundary layer is much thinner than the channel.) Two invariants are $y/x^{1/3}$ and $C/x^{\alpha/3}$, so we set

$$C = x^{\alpha/3} \, g\!\left(\frac{y}{x^{1/3}}\right). \qquad (23)$$

In order to satisfy (21a), α must equal 1, and $\dot{g}(0)$ must equal -1. We see at once, then, that the excess wall concentration rises as the 1/3-power of the distance down the channel. This conclusion is entirely a consequence of the group invariance since it did not involve solving any deqs.

Let us digress for a moment and rewrite this formula in ordinary units:

$$C = g(0) \left(\frac{xv^3}{u_y D^2}\right)^{1/3} C_0. \qquad (24)$$

All that is missing is the unknown constant $g(0)$, so Eq. (24) is very much like the results one obtains from dimensional analysis. But the reader should remember that (24) cannot be obtained from purely dimensional considerations — the fact of group invariance needs to be added. Equation (24) as it stands, that is, without knowledge of $g(0)$, can be used to correlate data taken for different salts, different permeation rates and different flow velocities, for example.

If we substitute (23) with $\alpha = 1$ into (20) we get

$$\ddot{g} + \frac{1}{3} \eta^2 \dot{g} - \frac{1}{3} \eta g = 0, \quad \eta = y/x^{1/3}. \qquad (25)$$

Again $g = \eta$ is a special solution of the linear Eq. (25), so we set $g = \eta f$ and obtain the following separable equation for f:

$$\eta\ddot{f} + (2 + \eta^3/3)\dot{f} = 0. \tag{26}$$

Then

$$g = \eta f = A\left(e^{-\eta^3/9} - \frac{\eta}{3}\int_\eta^\infty \eta e^{-\eta^3/9}\, d\eta\right) \tag{27}$$

where A is a constant of integration. Now

$$g(0) = A \tag{28a}$$

while

$$-1 = \dot{g}(0) = -\frac{A}{3}\int_0^\infty \eta e^{-\eta^3/9}\, d\eta \tag{28b}$$

so

$$g(0) = \left(\frac{1}{3}\int_0^\infty \eta e^{-\eta^3/9}\, d\eta\right)^{-1} = \frac{9^{1/3}}{\Gamma(\frac{2}{3})} = 1.536. \tag{28c}$$

4 Nonlinear diffusion

4.1 Boltzmann's problem : $C_t = (CC_z)_z$

Boltzmann considered diffusion with a concentration (or temperature) dependent diffusion coefficient [BO94]. When the diffusion coefficient is directly proportional to the concentration (temperature) itself, the diffusion equation becomes, in suitable special units,

$$C_t = (CC_z)_z. \tag{1}*$$

This nonlinear equation is invariant to the group of transformations

$$\begin{aligned} C' &= \lambda^\alpha C \\ t' &= \lambda^\beta t \quad 0 < \lambda < \infty \\ z' &= \lambda z \end{aligned} \tag{2a}$$

with

$$\alpha + \beta = 2. \tag{2b}$$

Two invariants are $C/t^{\alpha/\beta}$ and $z/t^{1/\beta}$, so following the steps in the last chapter we shall take

---
* Equation (1) arises in other physical phenomena besides heat or chemical diffusion. For example, consider the isothermal percolation of a perfect gas through a microporous medium as described by Darcy's law:

$$\frac{\partial \rho}{\partial t} + \frac{\partial}{\partial z}(\rho V) = 0 \quad \text{(continuity)}$$

$$\frac{\partial p}{\partial z} + V \frac{\mu}{\kappa} = 0 \quad \text{(Darcy's law)}$$

$$p = \rho RT \quad \text{(equation of state)}$$

where ρ is the density, p the pressure, V the percolation velocity, μ the viscosity, κ the permeability, T the absolute temperature and R the gas constant. In special units in which $RT = \mu/\kappa = 1$, these three equations reduce to (1) with $p = C$.

$$C = t^{\alpha/\beta}\, y\left(\frac{z}{t^{1/\beta}}\right). \tag{2c}$$

Let us begin with the problem of an instantaneous heat pulse in the plane $z = 0$ at time $t = 0$. As before, conservation of material (energy) requires

$$\int_{-\infty}^{+\infty} C\, dz = 1 \tag{3a}$$

so that $\alpha = -1$ and $\beta = 3$. In addition to (3a), we must satisfy the boundary and initial conditions

$$C(z,0) = 0 \quad z > 0 \tag{3b}$$
$$C(\infty,t) = 0 \quad t > 0. \tag{3c}$$

Then

$$C = t^{-1/3}\, y\left(\frac{z}{t^{1/3}}\right) \tag{4}$$

and

$$3(y\dot{y})^{\cdot} + y + x\dot{y} = 0, \quad x = z/t^{1/3}. \tag{5}$$

This equation can be integrated to give

$$3(y\dot{y}) + xy = \text{constant}. \tag{6}$$

Since $\dot{y}(0) = t^{2/3}\, C_z(0,t) = 0$ by symmetry, the constant in (6) must be zero. It follows at once that

$$y = \frac{x_0^2 - x^2}{6} \tag{7}$$

where x_0 is a constant of integration. It is determined from the requirement (3)

$$1 = \int_{-\infty}^{+\infty} C\, dz = \int_{-x_0}^{+x_0} y\, dx = \frac{2}{9} x_0^3. \tag{8}$$

The solution (7) is sketched in Fig. 6.

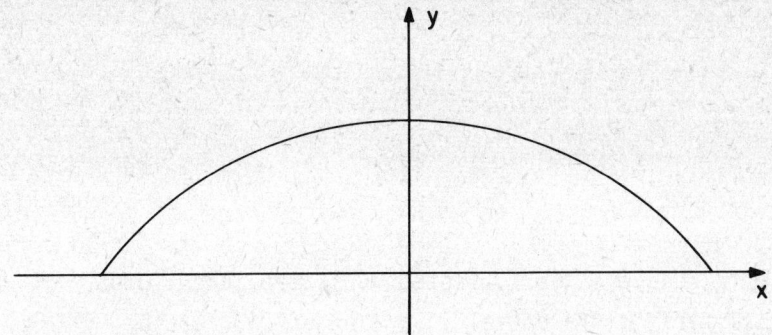

Fig. 6. A sketch of the solution (7)

The interesting thing about (7) is that $\dot{y}$ does not approach zero continuously as $x \to \infty$ as was the case for the linear diffusion. As before, (3b) and (3c) collapse to the same condition, namely $y(\infty) = 0$, but this is satisfied not just for infinite x, but for all $x > x_0$.

The solution (7) was first found by Pattle [PA59].

4.2 Clamped flux

Let us now consider the clamped flux case. The boundary condition (3a) is then replaced by

$$(CC_z)_{z=0} = \text{a constant, say, } -b. \tag{9}$$

Now $\alpha = 1/2$ and $\beta = 3/2$ so we set

$$C = t^{1/3} \, y\!\left(\frac{z}{t^{2/3}}\right) \tag{10}$$

and obtain for y

$$3(y\dot{y})^{\cdot} = y - 2x\dot{y}, \quad x = z/t^{2/3} \tag{11}$$

Condition (9) becomes

$$(y\dot{y})_{x=0} = -b \tag{12a}$$

while (3b) and (3c) collapse to

27

$$y(\infty) = 0. \qquad (12b)$$

The principal ode (11) is not readily integrable. It is, however, invariant to the associated group*

$$\begin{aligned} y' &= \mu^2 y \\ x' &= \mu x \end{aligned} \qquad 0 < \mu < \infty \qquad (13)$$

so we shall be able to reduce it to a first-order (associated) deq.

An invariant and a first differential invariant of (13) are $u = y/x^2$ and $v = \dot{y}/x$. A short calculation shows that in terms of u and v, (11) becomes

$$\frac{dv}{du} = \frac{u - 2v - 3v^2 - 3uv}{3u(v - 2u)} . \qquad (14)$$

Equation (14) is reasonably complicated and its direction field, shown in Fig. 7, is correspondingly complicated. The slope $dv/du = 0$ on the curve C:

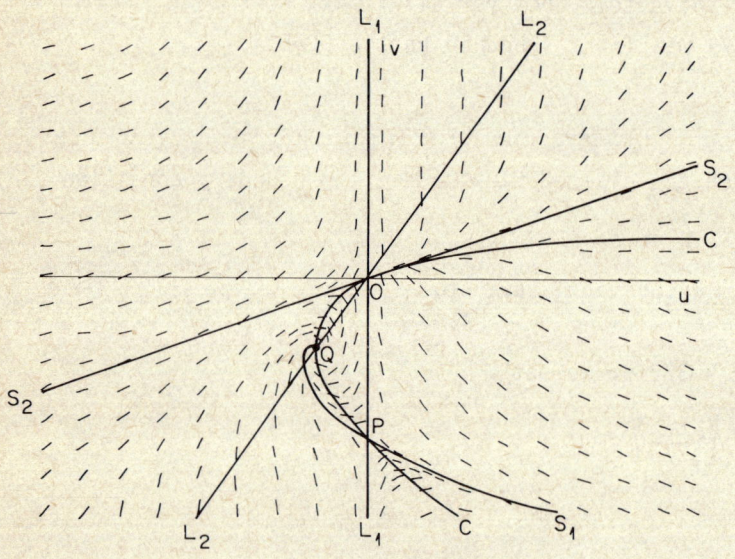

Fig. 7. The direction field of Eq. (14)

$u = v(2 + 3v)/(1 - 3v)$. (The curve C has two branches, one of which is shown

---
* Later in this chapter, we shall see when such associated groups exist and how they may be calculated independently of knowing the principal deq.

in Fig. 7. The other branch is in the second quadrant and is of no concern to us here since we are only interested in values of u > 0 and v < 0.) The slope dv/du = ∞ on the lines L_1:u = 0 and L_2:v = 2u. There are three singular points, O:(0,0), P:(0,-2/3), and Q:(-1/6, -1/3), and two separatrices, S_1 and S_2:v = u/2.

Following our experience with linear diffusion, we might begin by looking for a solution for which y and $\dot{y}$ → 0 as x → ∞. Such solutions must correspond to integral curves that pass through the origin of the (u,v)-plane. Since all integral curves entering the origin do so along S_2, no curve lying entirely in the fourth quadrant enters the origin. Furthermore, S_2:v = u/2 corresponds to the deq $y/2x^2 = \dot{y}/x$, which gives upon integration y = const. $\sqrt{x}$. So the curves entering the origin behave asymptotically as $\sqrt{x}$, which is unacceptable.

Thus, there are no solutions of (11) for which y and $\dot{y}$ approach zero continuously as x → ∞. This might have puzzled us if all we had to go on was the experience we had gained from linear diffusion problems. But now we know what to do: look for a solution that vanishes at a finite value of x, say x_0. Then at that point, u = $y(x_0)/x_0^2$ = 0 and v = $\dot{y}(x_0)/x_0$ < 0. Only one point can fill the bill, the singularity P, so the solution we are seeking corresponds to the separatrix S_1.

The value of v at P gives the slope of y at x_0: $\dot{y}(x_0) = -(2/3)x_0$. Knowing this, we can undertake a numerical integration of the second-order principal deq (11) to find y(x). The results of such an integration for the case x_0 = 1 are shown in Fig. 8. The value of $(y\dot{y})_{x=0}$ for this curve is 0.308. We can transform the solution of Fig. 8 according to the associated group (13) to obtain another solution having any desired value of $(y\dot{y})_{x=0}$. Now we have a complete solution to the problem originally stated at the cost of a single numerical integration of a second-order ode.

Even without undertaking this integration, we can learn useful things about the concentration (temperature) at the front face z = 0 from Eq. (14). When z → 0, x → 0 and u → ∞, v → -∞. So the behaviour of y near zero depends on the behaviour of the integral curves of (14) as we approach infinity through the fourth quadrant. Now when u >> 1 and |v| >> 1, v^2 >> |v| and u|v| >> u. So (14) can be written

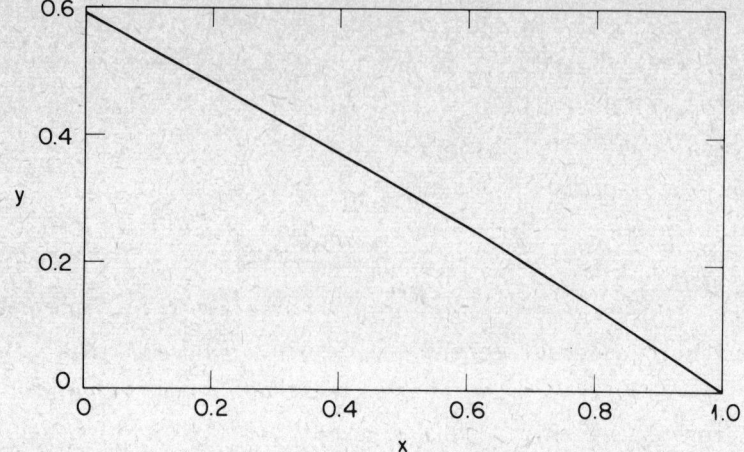

Fig. 8. The solution $y(x)$ to (11) for which $y(1) = 0$

$$\frac{dv}{du} = -\frac{v(v+u)}{u(v-2u)}. \qquad (15)$$

This limiting case of (14) is invariant to the transformation $u' = \lambda u$, $v' = \lambda v$ and so can be treated by the methods of Chapter 2. But even this is more than we need. We can further simplify (15) by noting from the direction field in Fig. 7 that on any integral curve in the fourth quadrant, $|v| \ll u$ for large enough u.* So (15) further simplifies to

$$\frac{dv}{du} = \frac{-vu}{u(-2u)} = \frac{v}{2u}. \qquad (16)$$

This integrates at once to give $v = -C\sqrt{u}$, where C is a (positive) constant of integration. If we substitute the values of u and v in terms of x and y, this last relation becomes $\dot{y} = -C\sqrt{y}$ so that

---

* Only three possibilities exist: $|v| \ll u$, $|v| \sim u$, or $|v| \gg u$. The first leads to (16). The third leads to $dv/du = -v/u$, which integrates to give $v = C/u$. But this contradicts the hypothesis $|v| \gg u$ as $u \to \infty$. The second alternative is equivalent to $v = au$, $a = $ a constant. Substituting this into (15) we find $a = -a(a+1)/(a-2)$ so that either $a = 0$ or $a = 1/2$, neither of which can be satisfied by an integral curve that approaches infinity in the fourth quadrant.

$$\frac{-\dot{y}(0)}{\sqrt{y(0)}} = C, \text{ a constant} \tag{17}$$

In terms of C, (17) becomes

$$C(0,t) = C^{-2/3} [-(CC_z)_{z=0}]^{2/3} t^{1/3} \tag{18a}$$

$$= C^{-2/3} b^{2/3} t^{1/3} \tag{18b}$$

Thus without solving any (but the most trivial) deqs we have discovered that the temperature at the front face at any time varies as the two-thirds power of the flux clamped at that face.

We can find the value of C by integrating the first-order deq (14) along the separatrix. Doing so we find $C = 0.679$; the same value follows from the solution to (11) shown in Fig. 8.

4.3 The associated group

The success we had in solving this last case of clamped flux depended on the existence of the associated group (13) to which the principal ode is invariant. The existence of this group is no accident, and the time has now come to explain the circumstances under which it exists. We consider pdes in one dependent and two independent variables, C, t, and z. We assume the pde is invariant to a *one-parameter family of stretching groups*

$$\begin{aligned} C' &= \lambda^\alpha C \\ t' &= \lambda^\beta t \quad 0 < \lambda < \infty \\ z' &= \lambda z \end{aligned} \tag{19a}$$

where the exponents α and β obey the linear constraint

$$M\alpha + N\beta = L \tag{19b}$$

The most general invariant relation connecting C, z, and t is then, by a now familiar argument,

$$C = t^{\alpha/\beta} y\left(\frac{z}{t^{1/\beta}}\right) \tag{20}$$

where y is an arbitrary function. The values of α and β are selected to satisfy the boundary conditions. Denote by α_0 and β_0 the particular values so selected. Then the solution we are seeking must have the form

$$C = t^{\alpha_0/\beta_0} y\left(\frac{z}{t^{1/\beta_0}}\right) = t^{\alpha_0/\beta_0} y(x), \quad x = \frac{z}{t^{1/\beta_0}} \tag{21}$$

Substitution of (21) into the pde yields an ode for $y(x)$.

If we transform $C(z,t)$ given by (21) according to transformations of family (19a) for which $\alpha \neq \alpha_0$ and $\beta \neq \beta_0$, its image $C'(z',t')$ must also satisfy the partial differential equation because the latter is invariant to all transformations (19a), not just those for which $\alpha = \alpha_0$ and $\beta = \beta_0$. Now

$$C'(z',t') = \lambda^\alpha C(z,t) \tag{22a}$$

$$= \lambda^\alpha C\left(\frac{z'}{\lambda}, \frac{t'}{\lambda^\beta}\right) \tag{22b}$$

$$= \lambda^\alpha \frac{(t')^{\alpha_0/\beta_0}}{\lambda^{\alpha_0\beta/\beta_0}} y\left[\frac{z'}{\lambda(t')^{1/\beta_0}} \lambda^{-\beta/\beta_0}\right] \tag{22c}$$

$$= \lambda^{(\alpha\beta_0-\alpha_0\beta)/\beta_0} (t')^{\alpha_0/\beta_0} y\left[\lambda^{(\beta-\beta_0)/\beta_0} \frac{z'}{(t')^{1/\beta_0}}\right] \tag{22d}$$

It is easy to verify that

$$\frac{\alpha\beta_0 - \alpha_0\beta}{\beta_0 - \beta} = \frac{L}{M} \tag{23}$$

as long as α, β and α_0, β_0 satisfy the linear constraint (19b). If we set $\mu = \lambda^{(\beta-\beta_0)/\beta_0}$, we can write (22d) as

$$C'(z',t') = (t')^{\alpha_0/\beta_0} \mu^{-L/M} y\left[\frac{\mu z'}{(t')^{1/\beta_0}}\right] \tag{24a}$$

or, dropping the primes, as

$$C(z,t) = t^{\alpha_0/\beta_0} \mu^{-L/M} y(\mu x). \tag{24b}$$

Because of the way it was obtained, the $C(z,t)$ of (24b) must be a solution of the pde. It, like the $C(z,t)$ of (21), is composed of a factor t^{α_0/β_0} times a function of $x = z/t^{1/\beta_0}$. Comparing (24b) with (21), we see that if $y(x)$ is such a function so is $\mu^{-L/M} y(\mu x)$. Now the function $\mu^{-L/M} y(\mu x)$ is the image of the function $y(x)$ under the transformation

$$y' = \mu^{L/M} y$$
$$x' = \mu x \qquad 0 < \mu < \infty \qquad (25)$$

For, $y'(x') = \mu^{L/M} y(x) = \mu^{L/M} y(x'/\mu)$, so if we replace μ by μ^{-1} we get $y'(x') = \mu^{-L/M} y(\mu x')$. So each function $y(x)$ satisfying the pde through Eq. (21) generates a one-parameter family of functions of x that do the same, namely, its images under the group of transformations (25). Each such *family* is invariant to the group (25).

Suppose the ode for $y(x)$ is of nth order. The solutions of such an equation form an n-parameter family of curves. From what we have just seen, this n-parameter family must decompose into an (n-1)-parameter set of one-parameter families, each of which is invariant to (25). But then the entire n-parameter family is invariant to (25). This means the ode for $y(x)$ is invariant to (25). The group (25) is the associated group of the principal family of groups (19).

The chief condition for the existence of an associated group is the invariance of the pde to a *one-parameter family of stretching groups*. This is a very high degree of symmetry and is only attained in the simplest of equations.

Had we known this result when we started the clamped-flux problem, we could have written down (17) and (18) without ever calculating the principal deq (11) or the associated deq (14). To see how this works, let us consider other boundary conditions corresponding to other physical problems.

4.4 Clamped concentration (temperature)

Suppose we clamp the concentration (temperature) on the front face at a fixed value starting at $t = 0$. What is the flux through the surface at $z = 0$? This particular problem has some practical interest, having been considered by Nilson in a study of the penetration of pressurized gases into a microporous half-space - see footnote to Equation (1). The clamped concentration boundary condition requires α to be zero, so $\beta = 2$. Then

$$C(z,t) = y\left(\frac{z}{t^{1/2}}\right) \tag{26a}$$

$$C(0,t) = y(0) \tag{26b}$$

and

$$C_z(0,t) = t^{-1/2} \dot{y}(0) \tag{26c}$$

The boundary and initial conditions that C must obey are (26b) and $C(z,0) = C(\infty,t) = 0$, which collapse to $y(\infty) = 0$. If we transform a solution $y(x)$ of the principal deq obeying the condition $y(\infty) = 0$ according to the associated group (13), we obtain another solution also obeying the condition $y(\infty) = 0$ but having the value $\mu^2 y(0)$ at $x = 0$. Thus all solutions $y(x)$ that we are seeking are images of one another under the associated group (13). Now, the quantity $\dot{y}(x)/\sqrt{y(x)}$ is invariant to (13). Since the point $x = 0$ is its own image under (13), the solutions we are seeking must all have the same value of $\dot{y}(0)/\sqrt{y(0)}$; call it $-A$. Thus

$$-C_z(0,t) = A \sqrt{C(0,t)} \cdot t^{-1/2} \tag{27}$$

where A is a constant not yet determined. According to (27), aside from its time dependence (already calculable from the principal group), the flux through the surface is proportional to the square root of the clamped (constant) surface concentration $C(0,t)$. The constant A can only be determined by studying the associated ode. The analysis is similar to that employed in connection with the clamped flux case; A turns out to be 0.452, which agrees within 1% with the value published by Nilson [NI81].

4.5 $C_t = (C^n C_z)_z$.*

By repeating this argument, we see that if the diffusion coefficient is proportional to C^n, then the flux and concentration at the front surface are related by

---

* Crank has pointed out that this equation describes diffusion of a species that can be strongly adsorbed according to a Freundlich isothermal [CR75]. The author has encountered it in the problem of the current distribution in superconductors undergoing a current ramp.

$$(-c^n c_z)_{z=0} \sim t^{-1/2} \cdot [C(0,t)]^{(2+n)/2} \qquad (28)$$

4.6 Exceptional solutions

The existence of the associated group has a further consequence that can sometimes be of importance, as we shall see subsequently. Following Section 7 of Chapter 2, we see that the associated ode must have the form

$$\frac{dv}{du} = \frac{G(u,v) - (\beta - 1)v}{v - \beta u} \qquad (2.37)$$

with $\beta = L/M$. If the simultaneous equations

$$G(u_0, v_0) = (\beta - 1)v_0 \qquad (29a)$$

$$v_0 = \beta u_0 \qquad (29b)$$

have a solution, it represents a singular point of (2.37) and corresponds to an exceptional solution $y = u_0 x^\beta = u_0 x^{L/M}$ of the principal ode. For, since $u = y/x^\beta$ and $v = \dot{y}/x^{\beta-1}$, the solution $y = u_0 x^\beta$ corresponds to $u = u_0$ and $v = \beta u_0 x^{\beta-1}/x^{\beta-1} = \beta u_0 = v_0$. Integral curves in the Lie plane that enter such a singularity correspond to solutions of the principal ode having $y = u_0 x^{L/M}$ as a limiting behaviour.

4.7 $CC_t = C_{zz}$.

The nonlinear diffusion equation

$$CC_t = C_{zz} \qquad (30)$$

occurs in the problem of the thermal expulsion of fluid from a long, slender, heated tube [DR81]. The quantity C represents the flow velocity induced in the fluid by the heating of the tube wall.* The boundary and initial conditions are

---

* In special units in which $Dc^2 = 4f$, where c is the sonic speed of the fluid (m s^{-1}), D is the diameter of the tube (m), and f is the Fanning friction factor.

$$C(z,0) = 0 \quad z > 0 \tag{31a}$$

$$C(\infty,t) = 0 \quad t > 0 \tag{31b}$$

and

$$C_z(0,t) = -b, \quad t > 0 \tag{31c}$$

where b is a constant related to the (constant) heating of the tube begun at $t = 0$.† The space coordinate z measures distance into the tube from the open end $z = 0$.

Equation (30) is also invariant to the family of groups (19a), but subject to the linear constraint

$$\alpha - \beta = -2 \tag{32}$$

in place of (19b). The boundary condition (31c) requires $\alpha = 1$, so $\beta = 3$. Thus

$$C(z,t) = t^{1/3} y\left(\frac{z}{\sqrt{3}\, t^{1/3}}\right) \tag{33}$$

where the ode for $y(x)$ ($x = z/\sqrt{3}\, t^{1/3}$) is invariant to the associated group

$$\begin{aligned} y' &= \mu^{-2} y \\ x' &= \mu x \end{aligned} \quad 0 < \mu < \infty \tag{34}$$

and obeys the collapsed boundary conditions $\dot{y}(0) = -b\sqrt{3}$ and $y(\infty) = 0$. (The factor of $\sqrt{3}$ has been introduced for convenience.) Thus $-\dot{y}/y^{3/2}]_{x=0}$ is invariant to (34) and therefore the same for all solutions irrespective of the numerical value of $\dot{y}(0)$; call it A. Then

$$C(0,t) = t^{1/3} y(0) = 3^{1/3} A^{-2/3} t^{1/3} b^{2/3} \tag{35}$$

The physical interpretation of this result is that the velocity of thermal efflux is proportional to the one-third power of the elapsed time after the heating has begun and the two-thirds power of the heating rate.

---

† $b = \beta \dot{q}/c_p$, where β is the volume coefficient of thermal expansion (K^{-1}), $\dot{q}$ is the heating rate (W kg^{-1}), and c_p is the specific heat (J kg^{-1} K^{-1}).

In this case, we shall find and solve the associated first-order ode and calculate the numerical value of the constant A. This particular problem has been chosen for complete treatment because its detailed solution will reveal further interesting features of the method. The principal ode is

$$\ddot{y} = y(y - x\dot{y}) \tag{36a}$$

with the boundary conditions

$$\dot{y}(0) = -b\sqrt{3} \tag{36b}$$

and

$$y(\infty) = 0 \tag{36c}$$

The choice of a differential invariant u and first differential invariant v is not unique; a convenient choice is

$$u = x^2 y \tag{37a}$$

$$v = x^2(y - x\dot{y}) \tag{37b}$$

After a short calculation, we find the associated ode

$$\frac{dv}{du} = \frac{v(2 - u)}{3u - v} \tag{38}$$

Figure 9 shows the Lie plane of Eq. (38). The slope vanishes on the lines L_1: $v = 0$ and L_2: $u = 2$ and is infinite on the line L_3: $v = 3u$. Accordingly, there are two singular points, the origin O: (0,0) and the point P: (2,6). The singular point P is a saddle point, the origin O is a node. Traversing P are two separatrices S_1 and S_2. One of them, S_1, also passes through the origin. Figure 9 also shows some typical integral curves labeled $I_1 - I_8$.

The solution we are seeking of the principal deq (36a) is finite and has a finite derivative at the origin $x = 0$. So the integral curve in the Lie plane corresponding to it must pass through the origin. Of the integral curves passing through the origin, S_1 is the one we want; for in the neighbourhood of the singularity P it will have the limiting behaviour $y = 2x^{L/M} =$

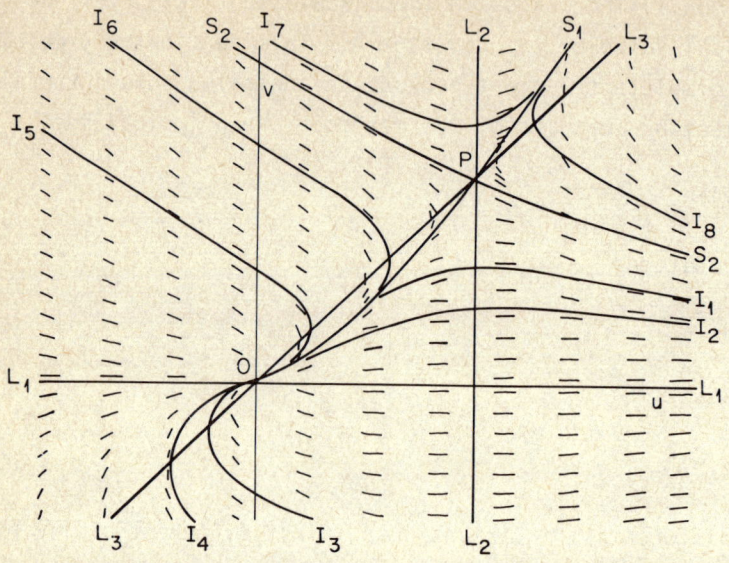

Fig. 9. The Lie plane of Eq. (38)

$2x^{-2}$, which will enable it to satisfy the boundary condition (36c).*

The situation we have here, namely, the sought-for integral curve in the Lie plane being a separatrix, is not at all uncommon, and will recur in other problems discussed in this book. Near the origin where $u \ll 2$, (38) becomes $dv/du = 2v/(3u - v)$, a homogeneous equation invariant to the stretching transformation $v' = \lambda v$, $u' = \lambda u$. So near the origin the separatrix must satisfy the algebraic equation

$$\frac{v}{u} = \frac{v(2-u)}{3u - v} \tag{39}$$

which says simply

$$v = u \tag{40}$$

---

* That no other integral curve passing through the origin can satisfy the boundary condition (36c) can be made plausible as follows. Consider the integral curves I_1 and I_2. When $u \to \infty$, $v \to 0$ as $e^{-u/3}$, as we can see by simplifying (38) in the limit. Now $v = 0$ for large x means $y - x\dot{y} = 0$ or $y = \text{const} \cdot x$. This can never obey (36c). This conclusion is satisfying, but if we believe the deq and boundary conditions specify a unique solution the identification of S_1 as allowing the boundary conditions to be met should be enough.

If we substitute $v = u + w$ into (38), then to lowest order we find $dw/du = 3w/2u$ so that $w = Cu^{3/2}$, where C is a constant of integration. This suggests that v may be expanded in powers of $u^{1/2}$ near the origin. If we set

$$v = u + Cu^{3/2} + Du^2 + Eu^{5/2} + \ldots \qquad (41)$$

substitute into (38), clear fractions and equate equal powers of u, we get

$$D = \frac{3}{2} C^2 - 1, \quad E = \frac{C}{4}(7D - 2), \ldots \qquad (42)$$

i.e., we get a series in which all the higher coefficients are determined by the value of the coefficient C. It is easy to see that

$$C = \lim_{u \to 0} \left(\frac{v - u}{u^{3/2}}\right) = - \frac{\dot{y}(0)}{[y(0)]^{3/2}} = A \qquad (43)$$

A similar procedure near the singular point P: (2,6) gives for the separatrix S_1

$$v = 6 + A(u - 2) + B(u - 2)^2 + \ldots \qquad (44)$$

where

$$A = (3 + \sqrt{33})/2, \quad B = A/(3A - 6), \ldots \qquad (45)$$

We can find the value of C on S_1 by using (44) to advance a short distance along S_1 from P. Then we integrate (38) numerically, advancing along S_1 towards O. When we get close to O, we match the numerical solution to the series (41) by choosing C correctly. In this way, with a single numerical integration, we find

$$C = 0.932 \qquad (46)$$

The numerical integration was carried out in the direction P → O because that direction of integration is stable. A glance at the direction field of Fig. 9 shows that neighbouring integral curves converge on S_1 as we move from P → O. So small errors caused by roundoff will heal themselves if we integrate from P → O. On the other hand, if we attempt to integrate from O → P, eventually we shall be thrown off either to the right or the left. Even if this difficulty did not occur, determination of C by integration in the direction O → P would require trial and error, whereas determination of C by integration in the direction P → 0 does not.

Another stability problem occurs near both singular points that has nothing to do with the direction of integration. Near each singularity, lines on which dv/du is either 0 or ∞ approach very closely to each other. A small error that puts us only slightly off the separatrix may cause a big change in the slope. This means that in the next integration step we shall step in a severely wrong direction. This causes another big change in slope and another step in the wrong direction. The net result is an erratic movement of the point (u,v) as the integration proceeds. To avoid this we must use the series (41) and (44) to stay away from the singularities. In the calculations leading to (46), I found it necessary to use five terms in series (44) and nine terms in series (41). Nonetheless, the calculations were relatively simple.

Once the value of C is in hand, calculation of y(x) by numerical integration of (36a) is an easy matter because consistent initial values of y(0) and ẏ(0) can be obtained from (43). Figure 10 shows the curve of y(x) for which

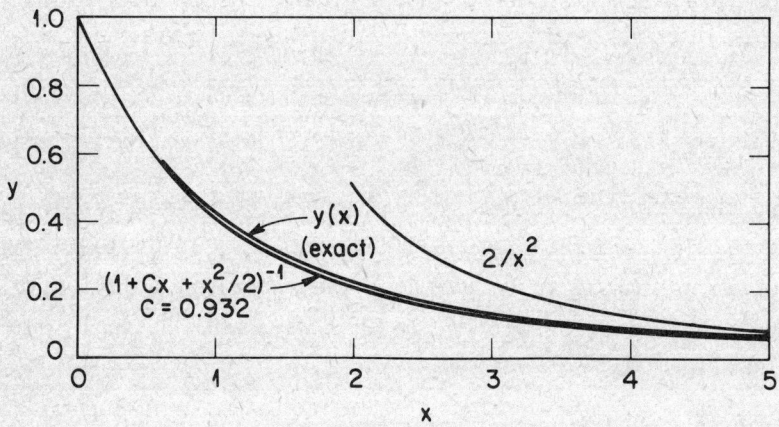

Fig. 10. The solution y(x) of (36a) for which y(0) = 1 and y(∞) = 0

y(0) = 1. As expected, the curve approaches $2/x^2$ for large x. This is fortunate because integrating in the direction of increasing x is the same as integrating in the direction 0 → P, so eventually the integration must become unstable. But we can approach the asymptotic limit $2/x^2$ closely enough that there is no difficulty in graphically continuing the solution. It is worth noting that the simple interpolation formula

$$y = (1 + Cx + x^2/2)^{-1} \tag{47}$$

gives quite a good fit to the results of the numerical integration.

4.8 $c^n C_t = C_{zz}$.

The clamped-flux problem can be solved for any value of n. From purely group-theoretic considerations we can show that

$$C(0,t) = A_n t^{1/(n+2)} |C_z(0,t)|^{2/(n+2)} \tag{48}$$

From the linear problem (n=0) worked in Chapter 3, we find that A_0 = 1.128. From the results of the problem just completed we find A_1 = 1.512. The author also solved the n = 2 case using the power-series method* just described and found A_2 = 1.604.

The author has made a practical application of (48) to transient heat transfer in near-critical single-phase helium. The specific heat of the fluid varies strongly with temperature near the pseudo-critical line and can be fitted, over the limited range of temperature rise of interest in the problem, by a value of n = 0.45. The corresponding value of A_n can be found by three-point interpolation from the values just quoted.

4.9 Transient heat transfer in superfluid helium

Helium has a low-temperature liquid phase (called He-II or superfluid helium) with rather unusual properties. One of the most unusual is that heat transport in stationary He-II is described not by Fourier's linear law, but by the nonlinear Gorter-Mellink law

$$q = -k\left(\frac{\partial T}{\partial z}\right)^{1/3} \tag{49}$$

Here q is the heat flux (Wm^{-2}), k is a kind of thermal conductivity ($Wm^{-5/3}K^{-1/3}$), and $\partial T/\partial z$ is the temperature gradient (Km^{-1}). Combined with the heat balance equation $S(\partial T/\partial t) + (\partial q/\partial z) = 0$ and in suitable special

---

* The radii of convergence of the two series is great enough in the n = 2 case to allow A_2 to be determined directly from the overlapping series without any need for a numerical integration.

units,* (49) leads to the nonlinear heat diffusion equation

$$\frac{\partial T}{\partial t} = \frac{\partial}{\partial z}\left[\left(\frac{\partial T}{\partial z}\right)^{1/3}\right] \tag{50}$$

This equation, too, is invariant to the family of groups (19a) but subject to the linear constraint

$$2\alpha - 3\beta = -4 \tag{51}$$

Invariant solutions of (50) take the form

$$T = t^{\alpha/\beta} y\left(\frac{z}{t^{1/\beta}}\right) \tag{52}$$

where $y(x)$ is a function whose ode is invariant to the associated group

$$\begin{aligned} y' &= \mu^{-2} y \\ x' &= \mu x \end{aligned} \qquad 0 < \mu < \infty \tag{53}$$

The quantity $a = -y^{3/2}/\dot{y}]_{x=0}$ does not change under transformations of the associated group (53).

In the foregoing equations T should be thought of as the temperature *rise*.

4.10 Clamped-flux

The clamped-flux case is of particular interest because it has been studied experimentally by van Sciver in connection with the stability of superconducting magnets cooled with superfluid helium [SC79]. The boundary conditions for this case are

$$T_z(0,t) = -(q/k)^3 \qquad t > 0 \tag{54a}$$

$$T(z,0) = 0 \qquad z > 0 \tag{54b}$$

$$T(\infty,t) = 0 \qquad t > 0 \tag{54c}$$

---

* In the special units, $k/S = T_\lambda - T_b = 1$. S is the heat capacity per unit volume, T_b is the ambient helium temperature, and T_λ is the (higher) temperature at which the superfluid experiences a phase change and becomes ordinary liquid He.

and require $\alpha = 1$, $\beta = 2$. Then they collapse to

$$\dot{y}(0) = -(q/k)^3 \qquad (55a)$$

$$y(\infty) = 0 \qquad (55b)$$

It follows then, as before, that

$$T(0,t) = t^{1/2} y(0) = a^{2/3} (q/k)^2 t^{1/2} \qquad (56)$$

Also as before, the dependence of $T(0,t)$ on q/k follows from the existence of the associated group (53).

An aspect of the relationship (56) has been checked experimentally by van Sciver. He measured the time it took for the temperature to rise from T_b to T_λ. According to (56), this time should scale as q^{-4}, and this is what van Sciver observed. If we solve the principal ode we can find the value of a and consequently calculate the constant of proportionality between t and q^{-4}. The author has done this and found good agreement with van Sciver's measured values [DR82].

The principal ode corresponding to the solution (52) is

$$\beta \frac{d}{dx}\left(\frac{dy}{dx}\right)^{1/3} + x \frac{dy}{dx} - \alpha y = 0 \qquad (57)$$

and if we choose $u = xy^{1/2}$ and $v = x\dot{y}^{1/3}$ as an invariant and first differential invariant of the associated group (53), the associated ode is

$$\frac{dv}{du} = \frac{u(2\beta v - 2v^3 + 2\alpha u^2)}{2\beta u^2 + \beta v^3} \qquad (58)$$

When $\alpha = 1$ and $\beta = 2$ this becomes

$$\frac{dv}{du} = \frac{u(2v - v^3 + u^2)}{2u^2 + v^3} \qquad (59)$$

Figure 11 shows a sketch of the fourth quadrant of the Lie plane of (59). Only the fourth quadrant is shown because the temperature rise is positive ($u > 0$) and falls as we move away from the heated plate ($v < 0$). The slope dv/du vanishes on the v-axis and on the curve C_1: $u^2 = v^3 - 2v$; it is infinite on the curve C_2: $2u^2 + v^3 = 0$. The origin O and the point P: $(2\sqrt[4]{3}/3, -2/\sqrt{3})$ are singular points.

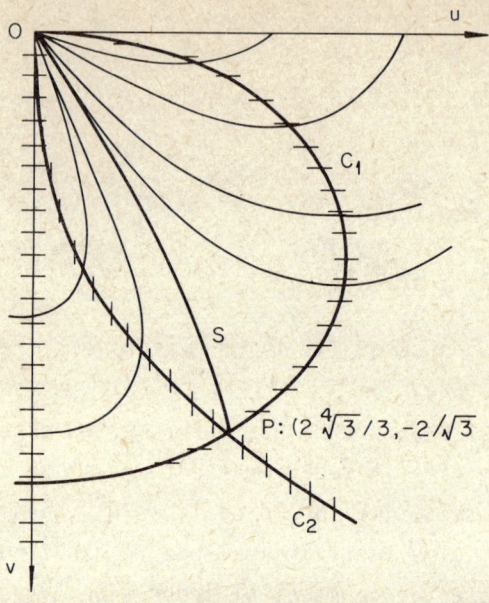

Fig. 11. Sketch of fourth quadrant of the Lie plane of (59)

When $x = 0$, u and v also equal 0, so we are interested in integral curves passing through the origin. Near the origin, one of the following three alternatives must hold, namely, (i) $v \ll u^2$, (ii) $v \sim u^2$, and (iii) $v \gg u^2$. In case (i) the deq (59) reduces to $dv/du = u/2$ which integrates to give $v = (u^2/4) + $ constant. This contradicts the hypothesis (i) which therefore cannot hold. Case (ii), $v \sim u^2$, means $v = bu^2$, where b is a constant. Substituting into (59) and keeping lowest-order terms we find $b = 1/2$. But this curve does not lie in the fourth quadrant, so we are left with alternative (iii). When $v \gg u^2$, (59) becomes $dv/du = 2uv/(2u^2 + v^3)$. (We cannot drop the v^3 term in the denominator because we do not know yet if $v^3 \ll u^2$, though subsequently we shall see that this is the case.) The last deq is invariant to the group $u' = \lambda u$, $v' = \lambda^{2/3} v$. Then we find $[v(u^2 - v^3)]^{-1}$ is an integrating factor, and a short computation gives for the integral $\Phi(u,v) = \ell n[(u^2 - v^3)/v^2]$. Thus, $u^2 = A^2 v^2 + v^3$, where A^2 is an arbitrary constant. When v is small enough, $A^2 v^2 \gg v^3$ so, close enough to the origin, $u = -Av$. Referring to the definitions of u and v, we see that $A = [-y^{3/2}(0)/\dot{y}(0)]^{1/3} = a^{1/3}$.

The integral curves that emanate from the origin are of two kinds, those that eventually intersect the curve C_1 and those that eventually intersect the curve C_2. The two kinds are separated by a separatrix S that joins the singularities O and P. The singular point P corresponds to the exceptional solution $y = 4\sqrt{3}/9x^2$, which should then also be the asymptotic behaviour of the solutions $y(x)$ corresponding to the separatrix S. This behaviour satisfies (55b), so again it is the separatrix we want.

The limiting behaviour $y = 4\sqrt{3}/9x^2$, when written in ordinary units is

$$\frac{T}{\sqrt{t}} = \frac{(4\sqrt{3}/9)(k/S)^{3/2}}{(z/\sqrt{t})^2} \tag{60}$$

This relationship is independent of a, so we need not perform any detailed calculations to compare it with experiment. The author has done this with van Sciver's points and found excellent agreement [DR82].

To find the separatrix, we must perform a numerical integration. The direction of stable integration is P → O. We use L'Hospital's rule to find the slope of S at P and advance a short distance from P towards O. Integrating towards O we then find without difficulty that A = 0.9132. Once we know A, we can find consistent initial conditions at x = 0. Then we can find the integral curves $y(x)$ by integrating the principal deq (57) away from x = 0. Because we integrate in the direction O → P, the integration eventually becomes unstable. But we can integrate close enough to the asymptotic limit $y = 4\sqrt{3}/9x^2$ that there is no difficulty in graphically continuing the numerical solution valid for small x. We need only do one integration for one consistent pair of values $y(0)$, $\dot{y}(0)$. The integral curves for other pairs of values can be obtained by transformation with the group (53). Figure 12 shows the integral curve for which $y(0) = 1$, $\dot{y}(0) = -1.313$.

4.11 Clamped temperature

This case is less interesting because no experimental data exist for it. It is solvable in simple terms and we quickly sketch the solution. The boundary condition (54a) is now replaced by the condition $T(0,t) = T_0$ for $t > 0$. To satisfy this condition, α must be zero, and therefore $\beta = 4/3$. Thus $T = y(z/t^{3/4})$. A point on the self-similar temperature profile marked by a particular temperature rise advances in a time t a distance z proportional to $t^{3/4}$. Furthermore,

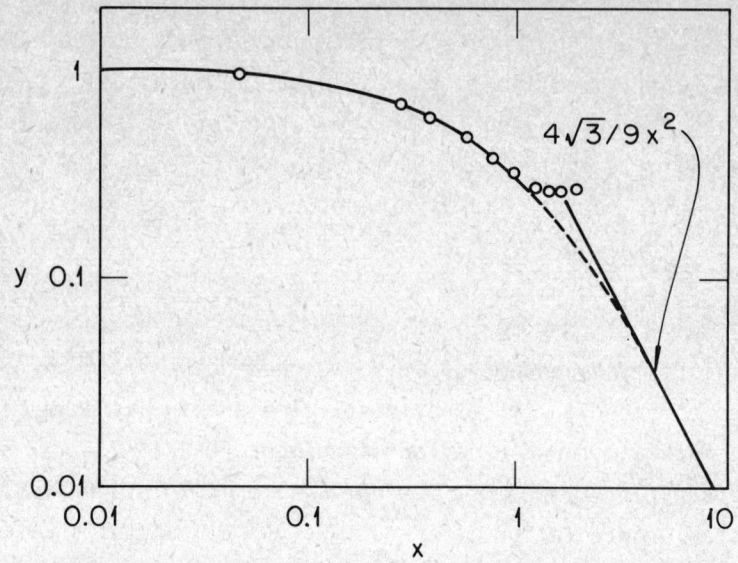

Fig. 12. The integral curve of (57) ($\alpha = 1$, $\beta = 2$) for which $y(0) = 1$. The circles have been obtained by a numerical integration that eventually becomes unstable. The dashed part of the curve is a graphical interpolation between the circles and the asymptote $4\sqrt{3}/9x^2$.

$$-T_z(0,t) = t^{-3/4} T_0^{3/2}/a \qquad (61)$$

where, as before, $a = -y^{3/2}/\dot{y}]_{x=0}$. So the heat flux through the surface is proportional to the $-3/4$-power of the elapsed time and the $3/2$-power of the clamped temperature rise.

When $\alpha = 0$, (57) can be solved easily by introducing $\dot{y}^{1/3}$ as a new dependent variable. The solution for which $T_0 = 1$ is

$$y = 1 - \frac{x}{\left(\frac{8}{3\sqrt{3}} + x^2\right)^{1/2}} \qquad (62)$$

Solutions for other values of T_0 can be found by transformation with (53). From (62) it follows that $a = (8/3\sqrt{3})^{1/2} = 1.241$. Solution (62) has the asymptotic behaviour $x^2 y = [2\beta/(2+\alpha)]^{3/2}/2$ that arises from the singular point P: $v_P = -[2\beta/(2+\alpha)]^{1/2}$, $u_P^2 = -v_P^3/2$.

4.12 Instantaneous heat pulse

This problem, too, is solvable in simple terms. Now the conservation condition (3a) replaces (54a), so $\alpha = -1$ and $\beta = 2/3$. We can integrate (57) at once to find

$$\frac{2}{3} \dot{y}^{1/3} + xy = 0 \tag{63}$$

the constant of integration being zero on account of (54b,c) which become (55b). From this we obtain by another integration

$$y = \frac{4}{3\sqrt{3}} (x^4 + b^4)^{-1/2} \tag{64}$$

where b is a constant determined from the integral condition $\int_{-\infty}^{+\infty} y\,dx = 1$. For other values of the integral, the corresponding solution can be determined by transformation with the group (53). For unit integral,

$$b = 2 \frac{\left[\Gamma\left(\frac{1}{4}\right)\right]^2}{3\sqrt{3\pi}} = 2.855 \tag{65}$$

Equation (64), too, has the asymptotic behaviour $x^2 y = [2\beta/(2+\alpha)]^{3/2}/2$ of the exceptional solution.

4.13 Isothermal percolation of turbulent liquid into a porous half-space

The next problem we shall deal with in this chapter is the isothermal percolation of a turbulent liquid into a porous half-space. The equations describing the fluid motion are those of continuity, motion and state:

$$\frac{\partial \rho}{\partial t} + \frac{\partial}{\partial z}(\rho V) = 0 \quad \text{(continuity)} \tag{66a}$$

$$\frac{\partial p}{\partial z} + \frac{2f\rho}{D} V^2 = 0 \quad \text{(motion)} \tag{66b}$$

$$\rho\left(\frac{\partial p}{\partial \rho}\right)_T = \text{constant (state)} \tag{66c}$$

Here ρ is the liquid density, V is the interstitial (pore) velocity, p is the pressure rise, D is the hydraulic diameter of the pores, f is the Fanning friction factor, and $\rho(\partial p/\partial \rho)_T$ is the bulk modulus of the liquid. In the equation of motion (66b), the acceleration term has been dropped because turbulent friction greatly dominates the inertial force. In other words, the pressure gradient is largely consumed in overcoming turbulent friction and not in accelerating the liquid. If we treat both the flow velocity and the changes in liquid density as small and keep terms only of lowest order, then in special units in which $\rho = (\partial p/\partial \rho)_T = D/2f = 1$, (66) become

$$\frac{\partial p}{\partial t} + \frac{\partial V}{\partial z} = 0 \qquad (67a)$$

$$\frac{\partial p}{\partial z} + V^2 = 0 \qquad (67b)$$

Either of the variables p or V may now be eliminated. If we eliminate V we get the pde $\partial p/\partial t = -\partial/\partial z(-\partial p/\partial z)^{1/2}$, which is similar to (50). If we eliminate p, we get the pde $2VV_t = V_{zz}$, which is similar to (30). However, we need not eliminate either variable but can deal directly with the system (67) of two simultaneous first-order pdes. Equations (67) are invariant to the transformations

$$\begin{aligned} V' &= \lambda^\gamma V \\ p' &= \lambda^\alpha p \\ t' &= \lambda^\beta t \\ z' &= \lambda z \end{aligned} \qquad 0 < \lambda < \infty \qquad (68a\text{-}d)$$

subject to the two linear constraints

$$\alpha - 2\beta = -3 \qquad (68e)$$

$$\gamma - \beta = -2 \qquad (68f)$$

The most general solution invariant to (68) must have the form

$$p = t^{\alpha/\beta} y\left(\frac{z}{t^{1/\beta}}\right) \qquad (69a)$$

$$V = t^{\gamma/\beta} w\left(\frac{z}{t^{1/\beta}}\right) \qquad (69b)$$

Substitution of (69) into (67) will result in a pair of coupled odes for the functions y and w. A repetition of the reasoning in Section 4.3 shows that these odes are invariant to the associated group

$$x' = \mu x$$
$$y' = \mu^{-3} y \qquad \text{(70a-c)}$$
$$w' = \mu^{-2} w$$

If we imagine the pressure rise on the exposed face of the half-space suddenly clamped at some value p_0, the boundary and initial conditions are then

$$p(0,t) = p_0 \quad t > 0$$
$$V(\infty,t) = p(\infty,t) = 0 \quad t > 0 \qquad \text{(71a-c)}$$
$$V(z,0) = p(z,0) = 0 \quad z > 0$$

Equation (71a) requires $\alpha = 0$, so that $\beta = 3/2$ and $\gamma = -1/2$. Now since solutions (69) corresponding to different values of p_0 are images of one another under the associated group (70), they all have the same value of $y(0)/[w(0)]^{3/2}$; call it B. Then,

$$V(0,t) = w(0) t^{-1/3} = B^{-2/3} p_0^{2/3} t^{-1/3} \qquad (72)$$

Thus the velocity of infiltration varies inversely as the 1/3-power of the elapsed time and directly as the 2/3-power of the clamped pressure rise at the front face. Equation (72), obtained entirely by group-theoretic means, constitutes a solution for the velocity of infiltration complete up to one as yet undetermined constant, B.

The coupled odes for y and w in case $\alpha = 0$ are

$$x\dot{y} = \frac{3}{2} \dot{w} \qquad \text{(73a)}$$
$$\dot{y} = -w^2 \qquad \text{(73b)}$$

and are easily solved to give

$$w = \left[\frac{1}{w(0)} + \frac{x^2}{3}\right]^{-1} \qquad \text{(74a)}$$

$$y = \int_x^\infty \frac{dx}{\left[\frac{1}{w(0)} + \frac{x^2}{3}\right]^2} \qquad \text{(74b)}$$

It follows from (74b) that

$$B = \frac{y(0)}{[w(0)]^{3/2}} = \sqrt{3} \int_0^\infty \frac{dx}{(1+x^2)^2} = \frac{\sqrt{3}\,\pi}{4} = 1.3603 \qquad (75)$$

Then $B^{-2/3} = 0.8145$; Nilson obtained the value 0.816 numerically [NI81].

4.14 Other groups

Barenblatt and Zeldovich [BA72] have studied Eq. (1) in a half-space with the boundary condition at the front face

$$C(0,t) = C_0 e^t \qquad -\infty < t < \infty \qquad (76)$$

Equation (1) is invariant to the one-parameter family of groups of transformations

$$\begin{aligned} C' &= e^{\alpha\lambda} C \\ t' &= t + \lambda \qquad -\infty < \lambda < \infty \\ z' &= e^{\alpha\lambda/2} z \end{aligned} \qquad (77)$$

where α is a constant that labels the member groups of the family. The boundary condition (76) is invariant to (77) when $\alpha = 1$.

The most general function $C(z,t)$ invariant to (77) must have the form

$$C = e^{\alpha t} y(z e^{-\alpha t/2}) \qquad (78a)$$

or, when $\alpha = 1$,

$$C = e^t y(z e^{-t/2}) \qquad (78b)$$

Here, as usual, y is an arbitrary function. If (78b) is a solution of (1), then its image under (77) must also be a solution of (1). In other words $C(z,t)$ given by

$$e^{\alpha\lambda} C = e^{t+\lambda} y(e^{\alpha\lambda/2} z e^{-(t+\lambda)/2}) \qquad (79)$$

should be a solution of (1), too. If we set $e^{(\alpha-1)\lambda/2} = \mu$, then (79) can be written

$$C = e^t \mu^{-2} y(\mu x), \quad 0 < \mu < \infty, \quad x \equiv z e^{-t/2} \tag{80}$$

So if $y(x)$ gives a solution of (1), so does $\mu^{-2} y(\mu x)$. Thus if $y(x)$ gives a solution of (1), so does any image of it under a transformation of the associated group

$$\begin{aligned} y' &= \mu^2 y \\ x' &= \mu x \end{aligned} \tag{81}$$

It follows from (78b) that

$$C_z(0,t) = e^{t/2} \dot{y}(0) \tag{82a}$$

$$C(0,t) = e^t y(0) \tag{82b}$$

so that

$$\frac{-C_z(0,t)}{\sqrt{C(0,t)}} = -\frac{\dot{y}(0)}{\sqrt{y(0)}} \tag{82c}$$

If the solutions $y(x)$ we are seeking are all images of one another under the associated group (81), the right-hand side of (82c) is simply a constant, A. Thus $-C_z(0,t) = A\sqrt{C_0}$, which means, in the case of Darcy-law infiltration (see footnote to Equation (1) above), that the velocity of infiltration at a fixed time scales as the square root of the clamped pressure.

To calculate the constant A, we must study the principal ode:

$$(y\dot{y})^{\cdot} = y - \frac{1}{2} x \dot{y} \tag{83}$$

As expected, it is invariant to the associated group (81). If we introduce the invariant and first differential invariant $u = y/x^2$ and $v = \dot{y}/x$, we find the associated, first-order ode

$$\frac{dv}{du} = \frac{2u - 2v^2 - v - 2uv}{2u(v - 2u)} \tag{84}$$

The Lie plane of (84) is exactly like that of (14) (see Fig. 7), and we therefore seek a solution of (83) that vanishes at some finite intercept x_0. There, $\dot{y}(x_0) = -x_0/2$ since there is only one singularity of (84) on the negative v-axis, namely, $(0,-1/2)$. With these *consistent* boundary values we can integrate (83) numerically to find $y(x)$. If all we want is A, we can integrate the first-order ode (84) out to large u, where it has the asymptotic form $v = -A\sqrt{u}$. (For large u and v, (84) reduces to (15), previously analyzed.) In order to advance away from the singularity on the negative y-axis we use L'Hospital's rule to find $(dv/du)_{u=0, v=-1/2} = -3/2$. Since the integral curves converge to the one we are seeking, a simple linear extrapolation is sufficient. The value of A so found is $A = 0.9075$.

4.15 Resemblance to dimensional analysis

The relationships (18), (27), (28), (35), (48), (56) and (72) resemble the results of dimensional analysis in that they connect products of powers of quantities of interest with an undetermined constant. They can be derived by a quick and easy procedure similar to that of dimensional analysis. The justification of the procedure, given below, is somewhat lengthier than the procedure itself.

Consider again a problem described by a pde invariant to the one-parameter family of groups (19). Suppose now we are interested in the relation between $C_z(0,t)$ and $C(0,t)$ when one or the other is clamped (or some product of powers of them is made proportional to a power of t, for that matter). For definiteness, let us take the case of clamped $C(0,t)$. Then

$$C_z(0,t) = F[C(0,t),t] \tag{85}$$

since the boundary value $C(0,t)$, the time t and the position z uniquely determine the entire solution $C(z,t)$ and its derivatives. If we transform the variables according to the group (19) we get a new problem with a new clamped boundary value $\lambda^\alpha C(0,t)$, but one which should be equally well described by the equation (85). So (85) should be invariant to (19). (Note that $z = 0$ transforms into $z' = 0$.) Thus

$$\lambda^{\alpha-1} C_z = F(\lambda^\alpha C, \lambda^\beta t) \quad 0 < \lambda < \infty \tag{86}$$

where for convenience we have stopped writing the arguments $(0,t)$ of C and C_z. By our standard procedure of differentiating with respect to λ and setting $\lambda = 1$, we find that the most general form C_z may have is

$$C_z = t^{(\alpha-1)/\beta} G\left(\frac{C}{t^{\alpha/\beta}}\right) \tag{87}$$

If we consider (87) written for one choice of α and β, say, $\alpha = \alpha_0$ and $\beta = \beta_0$, it should be invariant to the transformation (19) for some other value of α and β. So from

$$\lambda^{\alpha-1} C_z = t^{(\alpha_0-1)/\beta_0} \lambda^{\beta(\alpha_0-1)/\beta_0} G\left(\frac{\lambda^{\alpha} C}{\lambda^{\beta\alpha_0/\beta_0} t^{\alpha_0/\beta_0}}\right) \tag{88a}$$

and (87) we find

$$G(x) = \mu^{-1+M/L} G(\mu x) \tag{88b}$$

where x is an abbreviation for $Ct^{-\alpha_0/\beta_0}$ and μ is an abbreviation for $\lambda^{(\alpha\beta_0-\beta\alpha_0)/\beta_0}$. Differentiating (88b) with respect to μ and setting $\mu = 1$ immediately leads to the result

$$G(x) = \text{constant} \times x^{1-M/L} \tag{89}$$

Substituting (89) into (87), we find

$$C_z = \text{constant} \times C^{1-M/L} t^{-N/L} \tag{90}$$

Interestingly, the detailed character of the pde survives only through the coefficients M, N, and L of constraint.

If we review the procedure just followed, we see that we have used it to conclude that $C_z C^{-1+M/L} t^{N/L}$ is the only combination of these three variables that is invariant to (19) for all pairs of values of α and β constrained by (19b). We could more easily have found this invariant combination by transforming $C_z C^a t^b$ with (19):

$$C'_z C'^a t'^b = \lambda^{\alpha-1} C_z \cdot \lambda^{a\alpha} C^a \cdot \lambda^{b\beta} t^b \tag{91}$$

Thus

$$a\alpha + b\beta + \alpha - 1 = 0 \tag{92a}$$

Since α and β must satisfy

$$M\alpha + N\beta = L \tag{92b}$$

but are otherwise arbitrary, we find at once that

$$a = -1 + M/L, \quad b = N/L \tag{92c}$$

These last manipulations are exactly analogous to how one proceeds by dimensional analysis.

5 Boundary-layer problems

5.1 Prandtl-Blasius problem of a flat plate

The celebrated Prandtl boundary-layer equations for a flat plate are

$$uu_x + vu_y = \nu u_{yy} \tag{1a}$$

$$u_x + v_y = 0 \tag{1b}$$

in the coordinate system shown in Fig. 13. Here ν is the kinematic viscosity.

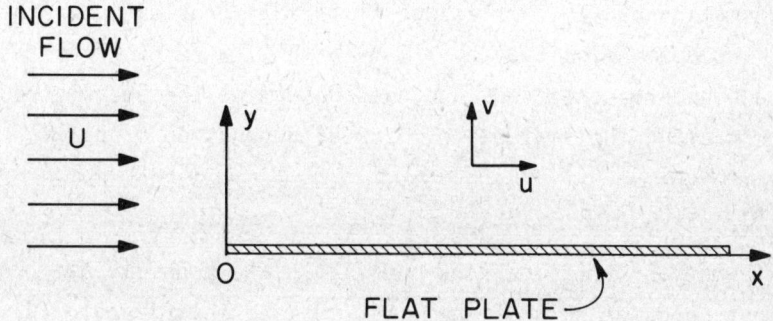

Fig. 13. Coordinate system used to represent the Prandtl boundary-layer equations. The origin O is at the leading edge of the flat plate. The incident flow is parallel to the flat plate with velocity vector $(u,v) = (U,0)$

The goal in the problem of the flat plate is the shear at the plate, $u_y(x,0)$. Knowing it, we can easily calculate the viscous drag on the plate.

Eqs (1a,b) are invariant to the groups of stretching transformations

$$\begin{aligned} u' &= \lambda^\alpha u \\ v' &= \lambda^{-1} v \\ x' &= \lambda^\beta x \\ y' &= \lambda y \end{aligned} \tag{2a}$$

55

where

$$\alpha - \beta = -2 \tag{2b}$$

The boundary conditions for the problem of the flat plate are

$$u = 0 \text{ at } y = 0, x > 0 \tag{3a}$$

$$v = 0 \text{ at } y = 0, x > 0 \tag{3b}$$

$$u = U \text{ at } y = \infty, x > 0 \tag{3c}$$

$$u = U \text{ at } x = 0, y > 0 \tag{3d}$$

These equations are all invariant to (2) when $\alpha = 0$ and $\beta = 2$. But even when $\alpha \ne 0$, eqs (3a,b) are invariant to (2) while (3c) and (3d) go into $u' = \lambda^\alpha U$. Thus transformation by (2) carries one flat-plate problem into another with a different incident velocity.

Since the boundary conditions and pdes determine the shear stress uniquely at every point along the flat plate, $u_y(x,0)$ depends on U and x:

$$u_y(x,0) = F(U,x) \tag{4}$$

Since transformation by (2) carries one flat-plate problem into another, and since (4) must hold for both of them, (4) must be invariant to (2) no matter what the value of α. A short calculation just like that in the previous section, 4.15, shows that

$$u_y(x,0) = \text{const.} \cdot U^{3/2} x^{-1/2} \tag{5a}$$

or in dimensionless form

$$\frac{\nu u_y(x,0)}{U^2} = \text{const.} \cdot \left(\frac{Ux}{\nu}\right)^{-1/2} \tag{5b}$$

The reader should note that (5b) cannot be determined by dimensional analysis alone. Pure dimensional analysis requires only that the two variables in (5b) be functionally related. Invariance to (2) must be invoked to show that the function is the square root. Equation (5b), which provides us with nearly all the information we require for practical purposes, is a pure group-theoretic consequence of the algebraic symmetry of the boundary-layer equations (1). All that is gained by completing the solution is the value of the constant.

5.2 Blasius's differential equation

To deal with (1a,b) it is convenient to introduce the stream function ψ defined by

$$u = \psi_y$$
$$v = -\psi_x \qquad (6)$$

and work in special units in which $\nu = U = 1$. The continuity equation (1b) is identically satisfied by (6), and (1a) becomes

$$\psi_y \psi_{xy} - \psi_x \psi_{yy} = \psi_{yyy} \qquad (7)$$

Under the transformation group (2), ψ transforms according to

$$\psi' = \lambda^{\alpha+1} \psi \qquad (8)$$

When $\alpha = 0$, the most general invariant form for ψ in terms of x and y is

$$\psi = \sqrt{x}\, f\!\left(\frac{y}{\sqrt{x}}\right) \qquad (9)$$

where f is a function yet to be determined. Then

$$u = \psi_y = \dot{f}(\eta), \quad \eta \equiv \frac{y}{\sqrt{x}} \qquad (10a)$$

$$-v = \psi_x = \frac{1}{2\sqrt{x}}(f - \eta\dot{f}) \qquad (10b)$$

and

$$2\dddot{f} + f\ddot{f} = 0 \qquad (11)$$

which is Blasius's differential equation. The boundary conditions (3) collapse to the three conditions

$$\dot{f}(0) = 0 \qquad (12a)$$

$$f(0) = 0 \tag{12b}$$

$$\dot{f}(\infty) = 1 \tag{12c}$$

which are just sufficient for the third-order equation (11). The shear at the plate is given by

$$u_y(x,0) = x^{-1/2}\ddot{f}(0) \tag{13}$$

so we must determine $\ddot{f}(0)$ in order to find the constants in eqs (5).

At first sight this procedure presents a small difficulty. Because the boundary conditions (12) are two-point boundary conditions, there is not sufficient information at $\eta = 0$ to undertake a numerical integration. Ordinarily in two-point boundary value problems we use the shooting method, which involves some trial and error. We guess a value of $\ddot{f}(0)$ and then integrate (11) numerically to large enough abscissas to allow an accurate estimate of $f(\infty)$. If it is not correct, we adjust the guessed value of $\ddot{f}(0)$ and repeat.

We can avoid this troublesome trial and error by exploiting the invariance of (11) to an associated group. The principal group of (7) is

$$\psi' = \lambda^{\alpha+1}\psi$$

$$x' = \lambda^{\beta}x \tag{14a}$$

$$y' = \lambda y$$

$$(\alpha + 1) - \beta = -1 \tag{14b}$$

The associated group is then

$$f' = \mu^{-1}f$$

$$\eta' = \mu\eta \tag{15}$$

to which (11) is clearly invariant. According to (15), $\dot{f}' = \mu^{-2}\dot{f}$ and $\ddot{f}' = \mu^{-3}\ddot{f}$; thus $\ddot{f}(0)/[\dot{f}(\infty)]^{3/2}$ is an invariant. The *first* numerical integration with any guess for $\ddot{f}(0)$ determines the value of this invariant, which is the same as that of the constant in eqs (5a) and (5b). The value is 0.33205.

5.3 The associated differential equation

In the present problem of boundary layer flow, a difficulty arises in the study of the associated differential equation that does not occur in the nonlinear diffusion problems dealt with in Chapter 4. Here, the principal deq is of *third* order. Invariance to the associated group (15) allows it to be reduced to a *pair* of coupled first-order equations. Whereas a single first-order equation can be studied conveniently by means of its direction field, a pair of coupled equations cannot so be studied.

In the problem at hand, however, this first difficulty can be circumvented. If the von Mises transformation is made in (1a) and (1b), i.e., if we pass from the independent variables x, y to new independent variables x, ψ, (1a) becomes the nonlinear diffusion equation

$$u_x = (u u_\psi)_\psi \tag{16}$$

which is of second order. It, too, is invariant to group (2). Its principal ode is of second order, too:

$$(u\dot{u})^\cdot + \frac{1}{2}\eta \dot{u} = 0, \quad \eta = \psi/\sqrt{x} \tag{17}$$

The shear $u_y(x,0)$ is given by $(u\dot{u})_{\eta=0}/\sqrt{x}$. If we write the transformations of group (2) relevant to (16) as

$$u' = \lambda^{\alpha/(\alpha+1)} u$$

$$x' = \lambda^{\beta/(\alpha+1)} x \tag{18a}$$

$$\psi' = \lambda \psi$$

and note that

$$\frac{\alpha}{\alpha + 1} + \frac{\beta}{\alpha + 1} = 2 \tag{18b}$$

we see that (17) should be invariant to the associated group

$$u' = \mu^2 u$$

$$\eta' = \mu \eta \tag{19}$$

which is easily verified directly. If we introduce the invariant $t = u/\eta^2$ and the first differential invariant $s = \dot{u}/\eta$, we obtain the associated deq

$$\frac{ds}{dt} = \frac{s(2s + 2t + 1)}{2t(2t - s)} \qquad (20)$$

Figure 14 shows the Lie plane of eq. (20). We are only interested in the first quadrant since both u and $\dot{u}$ are >0. The origin in the (s,t)-plane corresponds to the limit $\eta \to \infty$, since $\dot{u}(\infty) = 0$ and $u(\infty) = 1$. When s and t are both small, (20) becomes

$$\frac{ds}{dt} = \frac{s}{2t(2t - s)} \qquad (21)$$

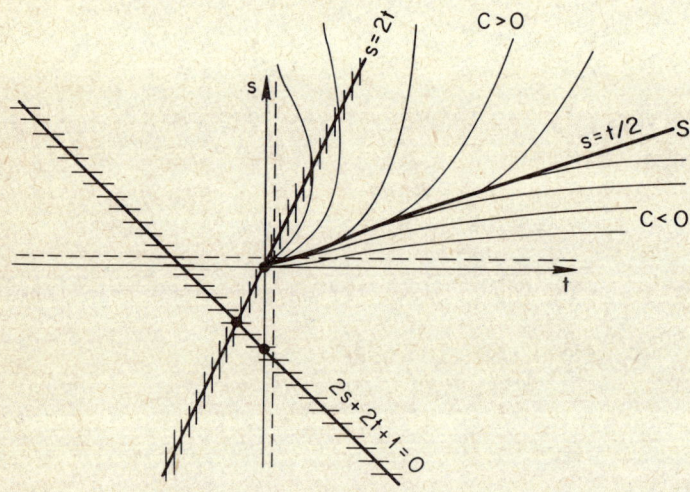

Fig. 14. The direction field of eq. (20)

How do the integral curves of (21) approach the origin? Can $t \gg s$ on such integral curves? If $t \gg s$, (21) becomes $ds/dt = s/4t^2$ which integrates to $s = \text{const} \times \exp(-1/4t)$. On these curves, t is indeed $\gg s$ when t is small. Can $t \sim s$, i.e., can $s = At$? Substitution of this hypothesis into (21) at once leads to a contradiction. Can $s \gg t$? Then (21) becomes $ds/dt = -1/2t$ which integrates to give $s = \text{const} + \ln(1/\sqrt{t})$. This function can never represent a curve passing through the origin. So only the first alternative is possible.

If we substitute for s and t in terms of u and $\dot{u}$, we then have for the first alternative

$$\frac{\dot{u}}{\eta} = \text{const} \times \exp\left(-\frac{\eta^2}{4u}\right) \tag{22}$$

when η is large. Since $u(\infty) = 1$, we can set $u = 1$ on the right-hand side and integrate again to find

$$u = 1 - \text{const.} \; e^{-\eta^2/4}, \quad \eta \gg 1 \tag{23}$$

When s and t are both large, (20) becomes

$$\frac{ds}{dt} = \frac{s(s+t)}{t(2t-s)} \tag{24}$$

Equation (24) is invariant to the stretching group $s' = \lambda s$, $t' = \lambda t$ and so an integrating factor can be found for it by Lie's method (Section 2.2). A short computation shows its integral curves to be represented by

$$s^4 t = C(2s - t)^3 \tag{25}$$

where C is a constant labeling the various integral curves. The family (25) represents two kinds of curves, those for which $C > 0$ and those for which $C < 0$. When $C > 0$, all points of an integral curve must lie above the line $s = t/2$. When $C < 0$ all points of an integral curve must lie below the line $s = t/2$. The line $s = t/2$ is the separatrix between these two families (a fact which can also be found directly by solving (24) with the method of Section 2.5).

The asymptotic form $s = t/2$ of the separatrix S between the two families of curves in Fig. 14 corresponds in terms of u and η to the differential equation $\dot{u}/\eta = u/2\eta^2$, which integrates to give $u = a\sqrt{\eta}$, where a is a constant of integration. Then $t = a/\eta^{3/2}$ and $s = a/2\eta^{3/2}$. Thus the remote part of S at large s and t gives the behaviour of u near $\eta = 0$. Finally, the constant $a = \sqrt{2(u\dot{u})}_{\eta=0}$.

The curves of the upper family $(C > 0)$ all intersect the line $s = 2t$ at $t = \sqrt{27C}/4$, which marks their maximum extent in t. Eventually, for each of them, $s \to \infty$ and $t \to 0$, conditions that are inconsistent with a constant value of $u\dot{u}$ at $\eta = 0$. The curves of the lower family, $(C < 0)$, are asymptotic for large s and t to $s = (-C)^{1/4}\sqrt{t}$ which leads to $u = \sqrt{-C}\eta^2/4$. This means that

61

$(u\dot{u})_{\eta=0} = 0$ for all the curves of this family, and they are unsuitable to represent $u(\eta)$.

The upshot of this analysis is that

$$u = a\sqrt{\eta} \qquad \eta \ll 1$$

$$= 1 - \text{const. } e^{-\eta^2/4} \qquad \eta \gg 1, \qquad (26)$$

which, though interesting, seems to bring us no closer to determining the value of a, our goal. I have gone through the details of the Lie plane analysis of Eq. (20) in order to show how one may use information like that contained in (26) to finish solving practical problems.

A function which (1) behaves like $\sqrt{\eta}$ for small η and (2) approaches 1 with a deviation proportional to $e^{-\eta^2/4}$ for large η is

$$u = \frac{\sqrt{2}}{\Gamma(1/4)} \int_0^\eta \frac{dq}{\sqrt{q}} e^{-q^2/4} \qquad (27)$$

For, the function defined by (27) varies as

$$u \sim \frac{2\sqrt{2}}{\Gamma(1/4)} \sqrt{\eta} \quad \text{for} \quad \eta \ll 1 \qquad (28a)$$

and

$$u \sim 1 - \frac{2\sqrt{2}}{\Gamma(1/4)} \eta^{-3/2} e^{-\eta^2/4} \quad \text{for} \quad \eta \gg 1 \qquad (28b)$$

Equation (28b) represents a somewhat faster approach to unity than required by (23), but the difference will only be felt at large η, when the deviation of u from 1 is insubstantial. According to (28a) $(u\dot{u})_{\eta=0} = [2/\Gamma(1/4)]^2 = 0.30430$, which is too low by about 9%. However, the derivative of a fitting function can be a much worse representation of the derivative of the fitted function than the representation of the function itself, so a better way to find $(u\dot{u})_{\eta=0}$ is to use the integral relation

$$(u\dot{u})_{\eta=0} = \frac{1}{2} \int_0^\infty \eta \dot{u} d\eta \qquad (29)$$

that follows from (17). From (29) we find that $(u\dot{u})_{\eta=0} = \Gamma(3/4)/\Gamma(1/4) = 0.33799$, which is too high by only 1.8% and adequate for practical purposes.

The reader should realize that the analysis of the last paragraph starting with eq. (27) is based entirely on guesswork.

5.4 Flat plate with uniform suction or injection

According to (10b), the only boundary condition on u at y = 0 other than (3b) that admits similarity solutions is $v(x,0) \sim x^{-1/2}$. So boundary-layer flow with *uniform* suction or injection cannot be described by similarity solutions. But the group (2) carries one case of uniform suction or injection into another, so the relation analogous to (4), namely

$$u_y(x,0) = F(U,x,v_w); \quad v_w = v(x,0) \tag{30}$$

should be invariant to (2). From the four variables in (30), two combinations invariant to (2) can be made, say, $u_y x^{1/2} U^{-3/2}$ and $v_w x^{1/2} U^{-1/2}$. Thus,

$$u_y(x,0) = U^{3/2} x^{-1/2} G\left(\frac{v_w x^{1/2}}{U^{1/2}}\right) \tag{31}$$

where G is an as yet undetermined function.

E.M. Sparrow, H. Quack and C.J. Boerner [SP70], have treated the problem of the flat plate with uniform suction or injection by changing in (7) to the variables

$$\xi = \frac{v_w x^{1/2}}{U^{1/2}} \tag{32a}$$

and

$$\eta = \frac{y U^{1/2}}{x^{1/2}}, \tag{32b}$$

both of which are invariant to (2). Then they substitute $\psi = \sqrt{x} f(\xi,\eta)$ into (7) obtaining (we pass again to special units in which $\nu = U = 1$)

$$2f_{\eta\eta\eta} + f f_{\eta\eta} = \xi(f_\eta f_{\xi\eta} - f_\xi f_{\eta\eta}) \tag{33}$$

The boundary conditions $u(\eta = 0) = 0$, $u(\eta = \infty) = 1$, and $v(\eta = 0) = v_w$ then become

$$f_\eta(\eta = 0) = 0 \tag{34a}$$

$$f_\eta(\eta = \infty) = 1 \tag{34b}$$

$$f + 2\xi = -\xi f_\xi \quad \text{at } \eta = 0 \tag{34c}$$

Sparrow et al. have tested the so-called method of local similarity on eqs (33) and (34). The basis of this method is to *assume* that the ξ-dependence of f is very weak. In this case, the right-hand sides of (33) and (34c) can be dropped; the term 2ξ on the left-hand side of (34c), however, is retained. Thus f becomes effectively a function of one variable, η, and ξ plays the role of a parameter. Equation (33) then becomes the ode (11), and the only difference from the Blasius problem is that (34c) replaces (12b).

The invariance of (11) to the associated group (15) can also be used in the method of local similarity to avoid the trial and error that otherwise would be involved in solving the two-point boundary-value problem we face in finding $f_{\eta\eta}(\eta = 0)$. If we add to (15) the stipulation that $\xi' = \mu^{-1}\xi$, then the equations of the local similarity theory are invariant. If we then fix ξ and guess a value of $f_{\eta\eta}(\eta = 0)$, we can integrate (11) and find $f_\eta(\eta = \infty)$. If this value is not 1, we scale it to 1 using (15), and scale $f_{\eta\eta}(0) \sim \mu^{-3}$ and $\xi \sim \mu^{-1}$ accordingly. Each integration of (11) gives us a pair of values $f_{\eta\eta}(\eta = 0)$ and ξ, and it is a relatively simple matter to construct a curve of $f_{\eta\eta}(\eta = 0)$ vs ξ.

If the ξ-dependent terms dropped were of smaller order of magnitude then the one that is kept, the method of local similarity would have some rigorous validity. One would expect not only the value but also the slope of the curve of $f_{\eta\eta}(\eta = 0)$ vs ξ to be correct at $\xi = 0$. Figure 15, based on the work of Sparrow et al. shows that this is not the case. So it would appear that the terms dropped are of the same order as those kept. For this problem, at least, then, the method of local similarity is not based on a consistent approximation.

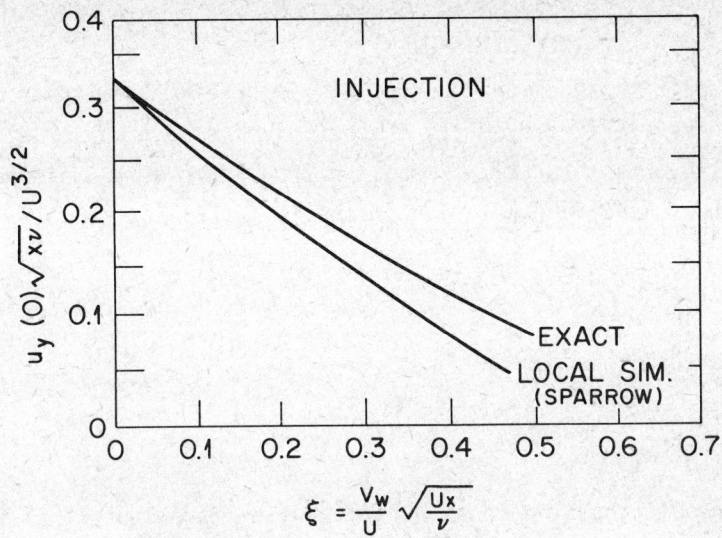

Fig. 15. The shear constant as a function of ξ according to the method of local similarity. The curves are taken from the paper of Sparrow et al. [SP70]

5.5 Thermal boundary layers

If the flat plate in Fig. 13 is heated, we must add to Eqs (1a,b) the equation for the *temperature rise* T

$$uT_x + vT_y = DT_{yy} \tag{35}$$

Here D is the thermal diffusivity. (In special units in which $\nu = 1$, it is numerically equal to the reciprocal of the Prandtl number.) Eq. (35) is invariant to group (2) as long as T' is linearly related to T (whether or not the coefficients of the linear relation depend on λ). Suppose now we consider the problem in which the plate is held at the same temperature T_w at every point. The local heat transfer coefficient $h \equiv -DT_y(y = 0)/T_w$* is independent of T_w because of the linearity of (35) and so can only depend on U and x.

---

* The special units can be augmented by choosing the unit of mass so that $\rho c_p = 1$.

$$\frac{h}{D} = F(U,x) \tag{36}$$

Transformation by (2) changes one boundary-layer problem (characterized by U) to another (characterized by U'). Since (36) must be true for all such problems, it is invariant to (2). Because h transforms as y^{-1}, it scales with λ^{-1}. The only invariant relation of the form (36) is then

$$\frac{h}{D} = \text{constant} \times \left(\frac{U}{x}\right)^{1/2} \tag{37a}$$

In dimensionless form this becomes

$$\frac{hx}{k} = \text{constant} \times \left(\frac{Ux}{\nu}\right)^{1/2} \tag{37b}$$

the well-known result that the local Nusselt number is proportional to the square root of the local Reynolds number.

Equation (37b) holds for any problem in which the temperature boundary condition is carried into a condition of the same type by group (2), e.g., clamped wall flux, wall temperature proportional to a power of x, wall flux proportional to a power of x, etc. Because of the linearity of (35), the "coefficient of proportionality" in the temperature boundary condition cannot enter (36), and the rest of the argument proceeds as before. Of course, the constant is different for different problems, but that is the only change.

To calculate the constant in the clamped-temperature case, we must remember that $\alpha = 0$ in the Blasius-Prandtl boundary layer. Thus we set $T = g(\eta)$ where $\eta = y/\sqrt{x}$. This function is the most general invariant to (2) that leaves the wall-temperature unchanged. A short calculation (again we use special units with $\nu = U = \rho c_p = 1$) then shows

$$\ddot{g} + \frac{f}{2D}\dot{g} = 0 \tag{38a}$$

with the boundary conditions

$$g(0) = T_w, \quad g(\infty) = 0 \tag{38b}$$

Here $f(\eta)$ is the solution of Blasius's deq (11). Equations (38) are easily integrable and give for the constant in (37b)

$$\frac{h\sqrt{x}}{D} = \frac{-T_y(0)\sqrt{x}}{T_w} = -\frac{\dot{g}(0)}{g(0)} = \left\{\int_0^\infty \exp\left[-\frac{Pr}{2}\int_0^\eta f(\eta')d\eta'\right]d\eta\right\}^{-1} \quad (39)$$

5.6 Free convection boundary layer

In this problem, the boundary-layer flow is induced by heat transferred from a *vertical* plate rather than being imposed from the outside. An additional term, βgT now appears on the right-hand side of (1a),* and in the boundary conditions (3c) and (3d) U is replaced by zero. Owing to the appearance of the βgT-term, T must transform like $T' = \lambda^{\alpha-2}T$ in order that (2) leave (1a) invariant. As before the local heat transfer coefficient h scales as λ^{-1}. Thus, by a now familiar procedure (treating βgT as a new variable in place of T) we obtain

$$\frac{h}{D} = \text{constant} \times \left(\frac{\beta gT_w}{x}\right)^{1/4} \quad (40a)$$

or in dimensionless form

$$\frac{hx}{k} = \text{constant} \times \left(\frac{\beta g x^3 T_w}{\nu^2}\right)^{1/4} \quad (40b)$$

i.e., the well-known result that the local Nusselt number is proportional to the one-quarter power of the Grashof number.

To find the velocity and temperature profiles when the plate temperature is clamped ($\alpha = 2$, $\beta = 4$), we set

$$\psi = x^{3/4} f\left(\frac{y}{x^{1/4}}\right) \quad (41a)$$

and

$$\beta gT = g\left(\frac{y}{x^{1/4}}\right) \quad (41b)$$

and find (in special units in which $\nu = 1$ and $\beta gT_w = 1$)

$$\dddot{f} + \frac{3}{4}f\ddot{f} - \frac{1}{2}\dot{f}^2 + g = 0 \quad (42a)$$

$$D\dot{g} + \frac{3}{4}f\dot{g} = 0 \quad (42b)$$

---

* β = volume coefficient of thermal expansion, g = acceleration due to gravity.

with the associated boundary conditions

$$f(0) = \dot{f}(0) = 0, \quad f(\infty) = 0$$

$$g(0) = 1, \qquad g(\infty) = 0 \tag{43}$$

These equations represent a difficult two-point boundary-value problem in the solution of which trial and error cannot be avoided. The reasons for this are worth considering for a moment. The boundary conditions (43) at infinity are zero boundary conditions, so we cannot alter them by scaling, in contrast to Blasius's case. So the fact that (42) are invariant to the associated group

$$\begin{aligned} f' &= \mu^{-1} f \\ g' &= \mu^{-4} g \\ \eta' &= \mu \eta \end{aligned} \tag{44}$$

is not of much help. If we guess $\ddot{f}(0)$ and $\dot{g}(0)$ wrongly, so that $f(\infty)$ and $g(\infty)$ do not vanish, scaling will not help because we cannot scale non-zero quantities to zero using (44).

In Chapter 4 on nonlinear diffusion we dealt with zero boundary conditions at infinity successfully. There we used invariance to the associated group to reduce the principal deq to a *first-order* associated deq. This equation could be analyzed by means of its Lie plane, and thus we could find the proper integral curve (a separatrix) to satisfy the zero boundary condition at infinity. However, the coupled deq (42) are of third and second order, respectively, so we cannot reduce their order enough with (44) to help much. In this problem, then, we seem to have passed limits beyond which the associated group cannot be used to avoid the trial and error inherent in two-point boundary problems.

6 Wave propagation problems

6.1 Introduction

The words "wave propagation" are used to characterize a variety of different phenomena, all involving the spreading of a disturbance in some medium. Conventionally, spreading of a disturbance by diffusion is not considered as wave propagation but, as we have seen in Sections 4.1 and 4.2, even diffusion problems can have solutions marked by sharp moving boundaries, one of the hallmarks of wave propagation. So we must beware of being too rigid in our classification of problems as wave propagation problems.

Most of the problems that will be dealt with in this chapter are in fact characterized by moving boundaries. The classical example of such a wave propagation problem having a similarity solution is a point explosion in a gas (Taylor [TA50], von Neumann [NE], Zeldovich [SE59]); the moving boundary in this case is a shock front. Gas-shock problems are mathematically complex because they are described by an equation of state and *three* coupled pdes, those of continuity, momentum conservation and energy conservation. A simpler system exhibiting the same mathematical phenomena is one considered by von Karman-Duwez [KA50] and Taylor [TA50], namely, compression and rarefaction waves in a nonlinearly elastic medium. Here the system is described by an equation of state and only *two* coupled pdes (continuity and momentum conservation). So we shall begin with the von Karman-Duwez-Taylor problem.

The wave motion in the examples just described can be termed longitudinal because the material motion is parallel to the direction of propagation of the moving boundary. A transverse wave is one in which the material motion is perpendicular to the direction of propagation of the moving boundary. An interesting transverse-wave problem having a similarity solution is that of a clamped membrane instantaneously loaded on one side by a pressurized gas (shock-loaded).

A classical moving-boundary problem based on diffusion as a mechanism is the Stefan problem: phase change (say, freezing) in a half-space induced by a sudden decrease in the temperature of the front face. Because this problem is not unlike those dealt with in Chapters 3 and 4, it will not be treated here in detail.

In all of the above problems the symmetry group is a stretching group. In problems invariant to translation groups, it is possible to have uniform wave propagation not characterized by a sharp boundary. The classical example is the solitary waves of hydrodynamics. Another example, dealt with here at modest length, is the spreading of a normal (non-superconducting) zone in a quenching superconductor.

These examples hardly exhaust the physical situations that lead to propagating-wave similarity solutions but their study should serve as a useful introduction to this interesting field.

6.2 von Karman-Duwez-Taylor problems

Let us consider a long rod of an elastic material (not necessarily Hookean) undergoing compression or rarefaction (Fig. 16). For simplicity, let us ignore lateral expansion or contraction, i.e., let us take Poisson's ratio to be zero. The left-hand shaded square represents an as yet undisturbed element of the rod with density ρ_0 lying between the Lagrangian coordinates

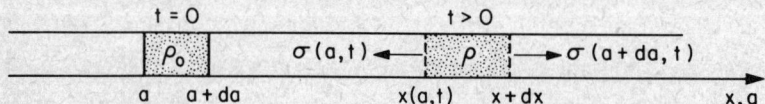

Fig. 16. Sketch to explain the notation used in von Karman-Duwez-Taylor problems

a and a + da. The right-hand shaded square represents the position of the same element later when it lies between the Eulerian coordinates $x(a,t)$ and $x + dx$. Its density is then $\rho(a,t)$, the *tensile* stress at its location $x(a,t)$ is $\sigma(a,t)$, and its velocity in the positive x-direction is $v(a,t)$.

Continuity requires that

$$\rho_0 da = \rho dx \tag{1}$$

The ratio $\rho_0/\rho = 1 + \eta$, where $\eta(a,t)$ is the local *tensile* strain. Thus

$$\frac{\partial x}{\partial a} = \frac{\rho_0}{\rho} = 1 + \eta \tag{2}$$

or, if we differentiate partially with respect to t,

$$\frac{\partial v}{\partial a} = \frac{\partial \eta}{\partial t} \tag{3}$$

The equation of motion of the element is

$$\rho_0 da \frac{\partial v}{\partial t} = \sigma(a + da) - \sigma(a) \tag{4a}$$

or

$$\frac{\partial v}{\partial t} = \left(\frac{1}{\rho_0} \frac{d\sigma}{d\eta}\right) \frac{\partial \eta}{\partial a} \tag{4b}$$

if we assume that the local tensile stress σ is a function only of the local tensile strain η. Once a constitutive equation relating σ to η (equation of state) has been chosen, (3) and (4b) can be used to calculate the evolution of any disturbance from its initial state.

6.3 Elastic (Hookean) wire

The problem we shall be considering in the next few sections is that of a thin elastic wire from the end of which a large weight is suspended (see Fig. 17). At $t = 0$, the weight is released and falls under the effect of gravity. We assume that the elastic restoring force of the wire is so small that the weight is freely accelerated downwards. What is the subsequent state of strain in the wire?

In this section we begin with the case of Hookean wire, i.e., one in which $\sigma = E\eta$, where E is Young's modulus. In special units in which $\rho_0 = E = 1$, (3) and (4b) become

$$\frac{\partial v}{\partial a} = \frac{\partial \eta}{\partial t} \tag{5a}$$

$$\frac{\partial v}{\partial t} = \frac{\partial \eta}{\partial a} \tag{5b}$$

The boundary conditions for the problem at hand are

$$v(0,t) = -gt \tag{6a}$$

$$v(\infty,t) = \eta(\infty,t) = 0 \tag{6b}$$

$$v(a,0) = \eta(a,0) = 0 \tag{6c}$$

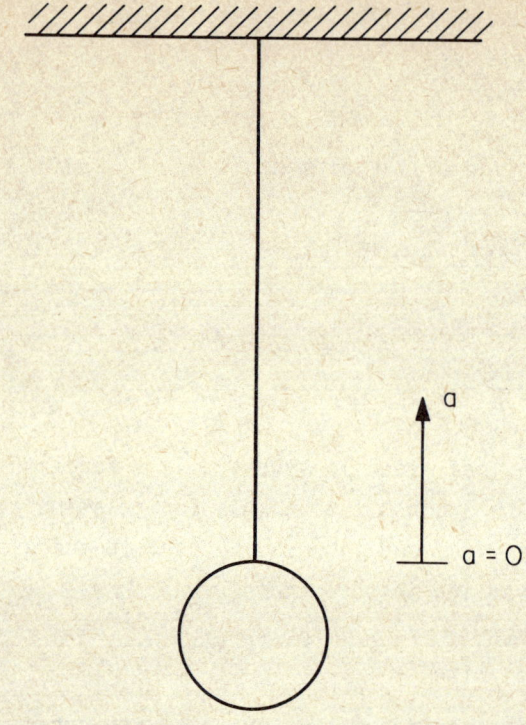

Fig. 17. Sketch of a large weight suspended from the end of a thin elastic wire. Shown also is the coordinate system being used

where g is the acceleration due to gravity, henceforth taken as unity (thus completely defining the special units). Eqs (5) are invariant to the group

$$v' = \lambda^\alpha v$$
$$\eta' = \lambda^\alpha \eta$$
$$t' = \lambda t \quad (7)$$
$$a' = \lambda a$$

and boundary condition (6a) requires $\alpha = 1$. Thus we take

$$v = tV\left(\frac{a}{t}\right) \quad (8a)$$

and

$$\eta = tH\left(\frac{a}{t}\right) \quad (8b)$$

In terms of V and H, (5) becomes

$$\dot{V} = H - x\dot{H} \tag{9a}$$

$$\dot{H} = V - x\dot{V} \tag{9b}$$

where $x \equiv a/t$, while Eqs (6) collapse to

$$V(0) = -1 \tag{10a}$$

$$V(\infty) = 0 \tag{10b}$$

$$H(\infty) = 0 \tag{10c}$$

If we add (9a) and (9b) we get

$$(\dot{V} + \dot{H})(1 + x) = V + H \tag{11a}$$

which integrates at once to give

$$V + H = \text{constant} \times (1 + x) \tag{11b}$$

In order to satisfy (10b) and (10c), the constant in Eq. (11b) must be zero, so $V = -H$. Then (9a) becomes

$$\dot{V} = -V + x\dot{V} \tag{12a}$$

which integrates to give

$$V = \text{const} \times (x - 1) \tag{12b}$$

Because of condition (10a), the constant in (12b) equals 1. So

$$\begin{aligned} V &= x - 1 & 0 \leqslant x \leqslant 1 \\ &= 0 & 1 \leqslant x \\ H &= 1 - x & 0 \leqslant x \leqslant 1 \\ &= 0 & 1 \leqslant x \end{aligned} \tag{13}$$

The lesson this simple example teaches us is the same lesson that we learned in Sections 4.1 and 4.2, namely that solutions occasionally satisfy the condition that they vanish at infinity by vanishing at some finite abscissa and remaining zero thereafter. The wave front $x = 1$ propagates uniformly with a velocity of $\sqrt{E/\rho_0}$ (in ordinary units). The maximum strain occurs at the end of the wire and is given by

$$\eta(0,t) = \frac{gt}{\sqrt{E/\rho_0}} \qquad (14)$$

in ordinary units.

6.4 Non-Hookean wire

If the wire in Fig. 17 is stretched beyond its elastic limit, the tension in it will no longer be a linear function of the strain, but rather will be a concave-downward function of the strain. We can try to simulate such behaviour by taking $\sigma = E\sqrt{\eta}$, in which case (3) and (4b) become

$$\frac{\partial v}{\partial a} = \frac{\partial \eta}{\partial t} \qquad (15a)$$

and

$$\frac{\partial v}{\partial t} = \frac{1}{\sqrt{\eta}} \frac{\partial \eta}{\partial a} \qquad (15b)$$

in special units in which $E/2\rho_0 = 1$. Eqs (15a,b) are invariant to the group

$$\begin{aligned} v' &= \lambda^\alpha v \\ \eta' &= \lambda^\gamma \eta \\ t' &= \lambda^\beta t \\ a' &= \lambda a \end{aligned} \qquad (16\text{a-d})$$

where

$$\alpha - 3\beta = -3 \qquad (16e)$$

and

$$\gamma - 4\beta = -4 \qquad (16f)$$

The boundary conditions are again those of Eq. (6) (special units: $g = 1$); because of Eq. (6a), α must equal β, so that $\alpha = \beta = 3/2$ and $\gamma = 2$. In this case, we can take

$$v = tV\left(\frac{a}{t^{2/3}}\right) \tag{17a}$$

$$\eta = t^{4/3}H\left(\frac{a}{t^{2/3}}\right)$$

as the most general invariant form. In view of the linear constraints (16e) and (16f) we expect that transforming by the associated group

$$x' = \mu x \quad (x \equiv a/t^{2/3})$$
$$V' = \mu^{-3}V \tag{18a-c}$$
$$H' = \mu^{-4}H$$

will carry one pair of acceptable functions V, H into another.

A short calculation shows that the principal odes for H and V are

$$\dot{V} = \frac{4}{3}H - \frac{2}{3}x\dot{H} \tag{19a}$$

$$\dot{H} = \sqrt{H}\left(V - \frac{2}{3}x\dot{V}\right) \tag{19b}$$

The boundary conditions (6) become

$$V(0) = -1 \tag{20a}$$
$$V(\infty) = 0 \tag{20b}$$
$$H(\infty) = 0 \tag{20c}$$

As anticipated, the principal odes (19) are invariant to the associated group (18).

It is easy to prove by a repetition of the argument of Section 2.7 that if we introduce two invariants of the associated group (18) in place of V and H in (19), then (19) reduce to a single first-order ode for one invariant in terms of the other. If we take, for example,

$$h = x^4 H \tag{21a}$$

and

$$w = x^3 V \tag{21b}$$

(19) becomes

$$\frac{dw}{dh} = \frac{3w - 2w\sqrt{h} + \frac{4}{3}h}{4h - \frac{8}{3}h^{3/2} + w\sqrt{h}} \tag{22}$$

Figure 18 shows a sketch of the fourth quadrant of the Lie plane of (22). (We are only interested in the fourth quadrant because $v < 0$, $\eta > 0$.) There

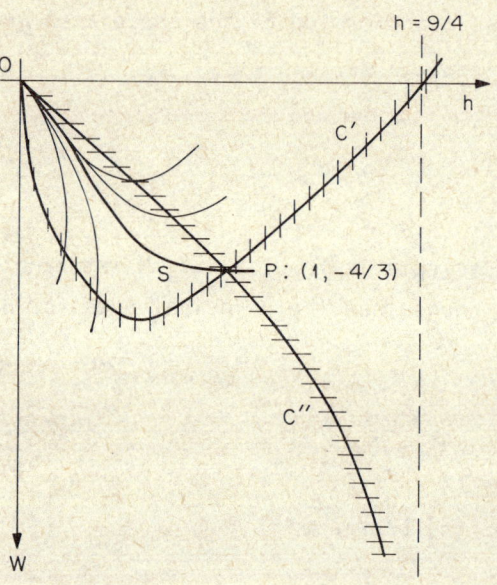

Fig. 18. Fourth quadrant of the Lie plane of Eq. (22)

are two singular points in the fourth quadrant, O, the origin, and P:(1,-4/3). The integral curves are of two types, those that cross C', the locus of infinite slope, and those that cross C", the locus of zero slope. These two types are separated by a separatrix S which passes through both singular points. In the fourth quadrant, all integral curves in the neighbourhood of the origin behave as $w = -Ch^{3/4}$. Since the origin O in the Lie plane corresponds to the point $x = 0$, the constant C is related to the value of H(0): $H(0) = C^{-4/3}$. The singular point P of the separatrix corresponds to the asymptotic behaviour $H \sim x^{-4}$, $V \sim -(4/3)x^{-3}$ and thus fulfills the boundary conditions (20b,c). If we take the problem to be physically well-determined, i.e., to have a unique solution, then it must be given by S. Now with a single numerical integration from $P \to O$ we can find C. Then we shall have consistent boundary conditions for the integration of (19), thereby solving our two-point boundary problem without any trial and error.

The particular manipulations involved in calculating C (using a series to advance away from P and then integrating numerically towards O and possibly joining the numerical solution to a series at O) can be tiresome; and since this problem has no special practical merit, we shall not perform them. What is interesting here we already know, namely, that the solution fills all space right from the start rather than being bounded by a sharp moving front, as in the preceding section. This happened, too, in the diffusion problems dealt with earlier, some being bounded and some not, but we made no attempt to relate this behaviour to the general structure of the pdes. But here we can do so, and a brief digression is worth while.

6.5 Characteristics and Riemann invariants

If we abbreviate $d\sigma/\rho_0 d\eta$ as c^2 and add c times eq. (3) to eq. (4b) we get

$$\left(\frac{\partial v}{\partial t} + c \frac{\partial v}{\partial a}\right) = c\left(\frac{\partial \eta}{\partial t} + c \frac{\partial \eta}{\partial a}\right) \tag{23}$$

The terms in parentheses are the directional derivatives of v and η, respectively, along the direction $da/dt = c$. So (23) says that the quantity $v - \int c d\eta$ is conserved along a curve whose slope da/dt equals $c(\eta)$ at each of its points. Similarly by subtracting the two equations, we find that $v + \int c d\eta$ is conserved along a curve whose slope $da/dt = -c$ at each of its points. The curves are called characteristics, and the quantities $v \pm \int c d\eta$ are called Riemann invariants.

In the case of a Hookean wire c is constant. Thus we can easily solve the problem of Section 6.3 by means of the wave diagram shown in Fig. 19. The axes of the wave diagram are a (abscissa) and t (ordinate). The lines of slope ±1 represent the characteristics (in special units in which $E = \rho_0 = c = 1$). Along characteristic AB, $v + \eta$ is constant and equal to zero (since $v = \eta = 0$ at point A). Thus $v = -\eta$ on AB and, as a matter of fact, on all negative-slope characteristics. So $v = -\eta$ everywhere. On the positive-slope characteristic, BC,

$$2v = v - \eta = v_B - \eta_B = 2v_B = -2t_B \tag{24}$$

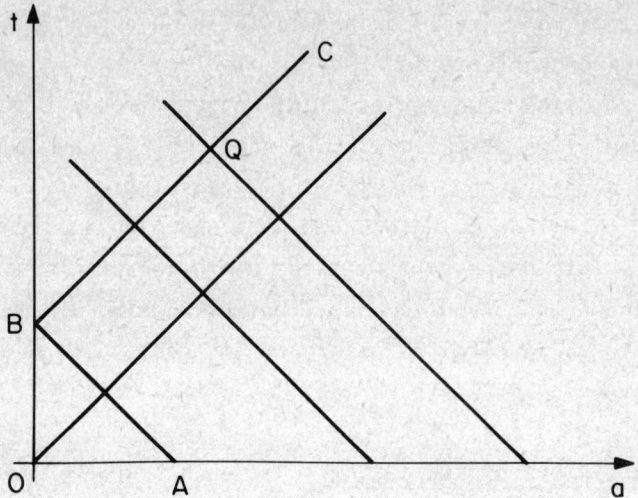

Fig. 19. Wave diagram for the Hookean wire

so for any point Q

$$v = -\eta = -t_B \tag{25}$$

where B is the intercept of the positive characteristic through Q on the line a = 0. With this established, it is easy to see that (8) and (13) give the proper solutions for v and η.

We proceed similarly for the non-Hookean wire of Section 6.4. In the special units we used there, $c = \eta^{-1/4}$. On all negative characteristics, therefore, $v + \frac{4}{3}\eta^{3/4} = 0$ so $v = -\frac{4}{3}\eta^{3/4}$ everywhere. Incidentally, from this we can see at once that

$$-C = \frac{w(0)}{h^{3/4}(0)} = \frac{V(0)}{H^{3/4}(0)} = \frac{v}{\eta^{3/4}}\Big]_{a=0} = -\frac{4}{3} \tag{26}$$

On positive characteristics the quantity $v - \frac{4}{3}\eta^{3/4} = -\frac{8}{3}\eta^{3/4}$ is conserved. Since η is then constant on each positive characteristic, so is $c = \eta^{-1/4}$. Thus the positive characteristics are straight lines!

On positive characteristics

$$-\frac{8}{3c^3} = -\frac{8}{3}\eta^{3/4} = v - \frac{4}{3}\eta^{3/4} = v_B - \frac{4}{3}\eta_B^{3/4} = 2v_B = -2t_B \tag{27}$$

so $c = (4/3t_B)^{1/3}$. Thus the equation of the positive characteristics is

$$a = \left(\frac{4}{3t_B}\right)^{1/3} (t - t_B) \text{ or } t = t_B + \left(\frac{3t_B}{4}\right)^{1/3} a \qquad (28)$$

Figure 20 shows the wave diagram for this problem with the positive characteristics sketched in. The smaller the intercept, the smaller the slope, so the

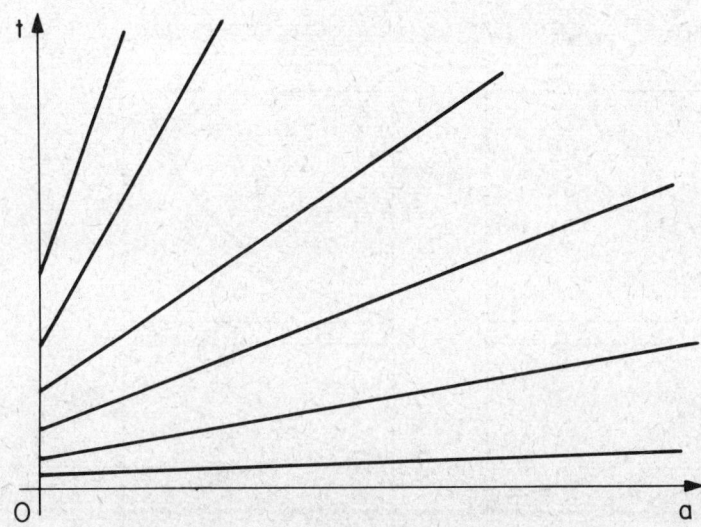

Fig. 20. Wave diagram for the non-Hookean wire showing the positive characteristics

positive characteristics fan out. In fact, the characteristics with infinitesimal t_B are nearly horizontal and shoot almost straight out to infinity. This is the reason why the disturbance fills all space right from the start.

We can see from the above argument that as long as c is a function only of η the positive characteristics will be straight lines. The behaviour of the family of straight lines depends on how c varies with η. If we repeat the argument given just above, we find that the equation of the positive characteristics can be written as

$$t = \frac{a}{c} + t_B \qquad (29a)$$

where c and t_B are related by

$$b = t_B \qquad (29b)$$

and the function b is defined by

$$b = \int_0^\eta c\,d\eta \qquad (29c)$$

What we need now is a plot of $c = db/d\eta$ versus b. If b is an increasing concave-downwards function of η (see Fig. 21a), then c will be a decreasing function of η and therefore of b, too. Thus, since $b = t_B$, c will be a decreasing function of t_B, and the positive characteristics will fan out.

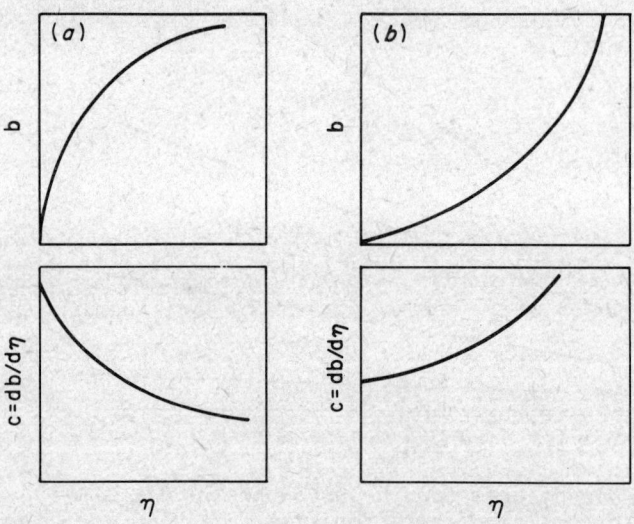

Fig. 21. Sketches of the behaviour of b and c as functions of η

What happens if b is an increasing, concave-*upwards* function of η (Fig. 21b)? Then c will be an increasing function of t_B, and the positive characteristics will look as shown in Fig. 22. Positive characteristics cannot intersect because then v and η are overdetermined. What happens then is that the positive characteristics are bounded by a moving shock front across which v and η undergo a discontinuous jump and no positive characteristic extends beyond its point of contact with the shock front. So now we must turn to consideration of such shock fronts.

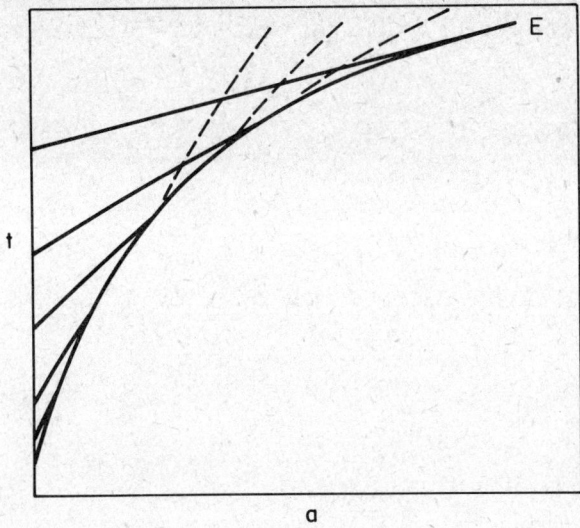

Fig. 22. Wave diagram in the case of intersecting characteristics

6.6 Shock conditions

The changes in strain and velocity across the shock front and its velocity of motion are not all independent but are constrained by the requirements of conservation of mass and momentum. Let us first consider these conservation conditions for a shock moving uniformly with velocity U in the laboratory system into an undisturbed medium (Fig. 23a). To simplify our calculations,

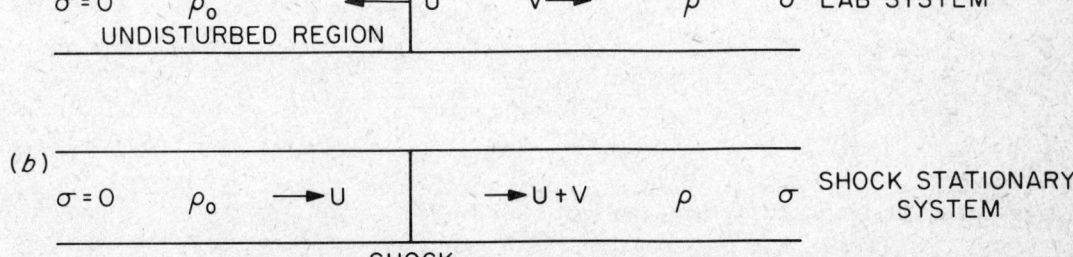

Fig. 23. Uniformly propagating shock in the lab and shock-stationary coordinate systems

let us consider the same shock in a coordinate system moving with the shock (Fig. 23b). In the shock-stationary system, conservation of mass requires

$$\rho_0 U = \rho(U + v) \tag{30a}$$

or

$$v = \eta U \tag{30b}$$

The momentum equation for the matter crossing the shock front in one second is

$$\rho_0 U v = \sigma(\eta) \tag{31a}$$

or

$$U^2 = \frac{\sigma(\eta)}{\rho_0 \eta} \tag{31b}$$

Eqs (31a) and (31b) are shock conditions analogous to the Rankine-Hugoniot conditions of gas dynamics.

6.7 "Superelastic" wire

Suppose now we consider a material which becomes stiffer as the strain increases, for example, a rubber band. For such a material, σ will be a concave-upwards, increasing function of η and so therefore will b. So such a material should show shock formation in our problem of the dropped weight.

Let us take for the sake of argument $\sigma = E\eta^2$ and use special units in which $E/\rho_0 = 1$. Then (3) and (4b) become

$$\frac{\partial v}{\partial a} = \frac{\partial \eta}{\partial t} \tag{32a}$$

$$\frac{\partial v}{\partial t} = 2\eta \frac{\partial \eta}{\partial a} \tag{32b}$$

The shock condition can be written

$$v^2 = \eta^3 \quad \text{(at the shock)} \tag{32c}$$

$$U^2 = \eta \tag{32d}$$

The deq (32a,b) are invariant to the group

$$v' = \lambda^\alpha v$$
$$\eta' = \lambda^\gamma \eta$$
$$t' = \lambda^\beta t \quad (33\text{a-d})$$
$$a' = \lambda a$$

with the linear constraints

$$\alpha + 3\beta = 3 \quad (33\text{e})$$
$$\gamma + 2\beta = 2 \quad (33\text{f})$$

The most general solution invariant to (33) is

$$v = t^{\alpha/\beta} V\left(\frac{a}{t^{1/\beta}}\right) \quad (34\text{a})$$

$$\eta = t^{\gamma/\beta} H\left(\frac{a}{t^{1/\beta}}\right) \quad (34\text{b})$$

The shock front must correspond to a fixed value A of $a/t^{1/\beta}$. Then the shock velocity will be $U \equiv da/dt = (A/\beta)t^{(1/\beta)-1}$. Now the shock conditions become

$$V^2(A) = H^3(A) \quad (35\text{a})$$

and

$$A^2 = \beta^2 H(A) \quad (35\text{b})$$

These relations are clearly invariant to group (33), too.

The linear constraints (33e,f) oblige the functions V and H to transform according to the associated group

$$V' = \mu^3 V \quad (36\text{a})$$
$$H' = \mu^2 H \quad (36\text{b})$$
$$x' = \mu x \quad (x \equiv a/t^{1/\beta}) \quad (36\text{c})$$

The boundary condition $v = -t$ at $a = 0$ (special units: $g = 1$) requires

$\alpha = \beta$, so $\alpha = \beta = 3/4$ and $\gamma = 1/2$. Then (34) become

$$v = tV\left(\frac{a}{t^{4/3}}\right) \qquad (37a)$$

$$\eta = t^{2/3} H\left(\frac{a}{t^{4/3}}\right) \qquad (37b)$$

Inserting these into (32), we find

$$\dot{V} = \frac{2}{3} H - \frac{4}{3} x \dot{H} \qquad (38a)$$

and

$$2H\dot{H} = V - \frac{4}{3} x \dot{V} \qquad (38b)$$

(which are invariant to the associated group (36) as expected). We must solve these equations subject to the boundary conditions

$$V(0) = -1 \qquad (39a)$$
$$V^2(A) = H^3(A) \qquad (39b)$$
$$16A^2 = 9H(A) \qquad (39c)$$

At this point, an easy procedure with which to finish this problem is to guess a value of A and calculate H(A) and V(A) using (39b) and (39c). Now we have sufficiently many boundary conditions at x = A to integrate inwards to the origin. Then we can transform the solution just found using the associated group (36) so that V(0) = -1. Figure 24 shows the results of such a calculation starting with A = 3, H = 16, V = 64. The value of V(0) obtained by numerical integration is -134.5. Scaling this to -1 with μ = 0.1952, we find A = 0.5855. The shock velocity is then

$$U = 0.7807 \left(\frac{Egt}{\rho_0}\right)^{1/3} \qquad (40)$$

in ordinary units.

As before, invariance of the principal odes to the associated group allows us to solve a two-point boundary problem without any trial and error.

Suppose we did not know that there was shock formation in this problem and we proceeded, as with previous problems, to try to find the "right" integral

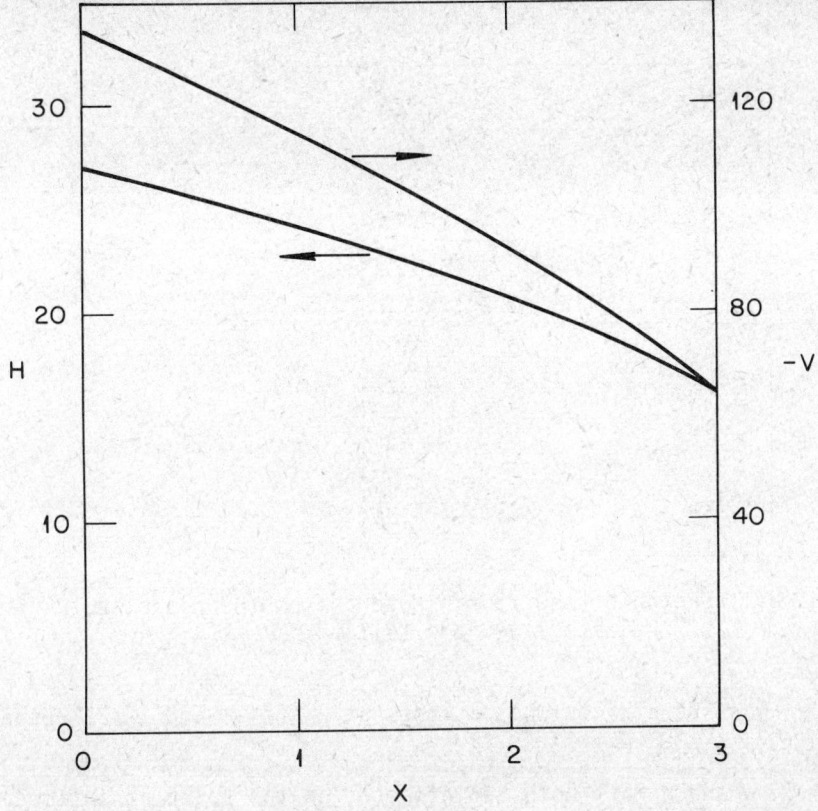

Fig. 24. Strain and velocity profiles behind the shock front

curve in the Lie plane. This attempt would have to fail (since no continuous solution exists). How exactly would it fail? The answer to this equation is interesting because if we can recognize the manner of failure, it may alert us to shock formation in other problems.

If we use the invariants $w = V/x^3$ and $h = H/x^2$ as new variables, (38) reduces to the associated ode

$$\frac{dw}{dh} = \frac{12w + 4h^2 - 18wh}{3w + 8h - 12h^2} \tag{41}$$

Since $V < 0$ and $H > 0$, we shall be interested in the fourth quadrant of the Lie plane. It is shown in Fig. 25. There are two singular points, one at the origin and one at P: (1/2, -1/3). The point P is a saddle point and is

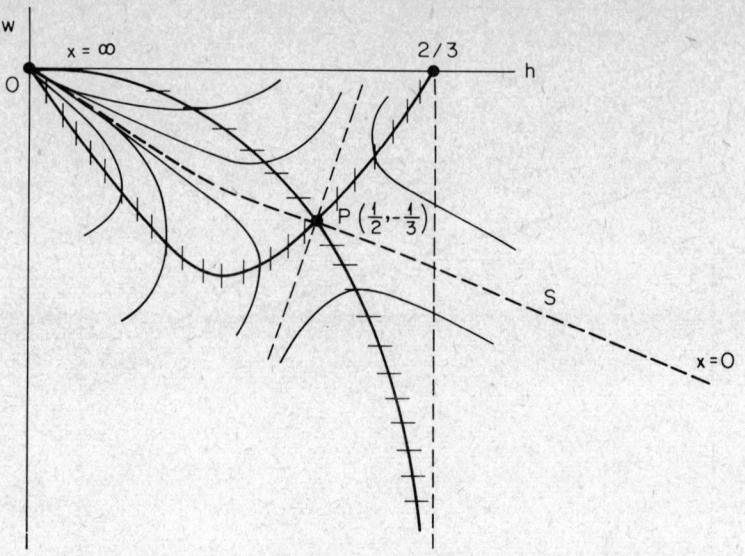

Fig. 25. Sketch of the fourth quadrant of the Lie plane of the associated ode (41)

traversed by two separatrices whose slopes at P are -1 and 10/3, respectively.

When $x = 0$, w and h must both be infinite, so the point at infinity in the Lie plane corresponds to the origin of x. Furthermore, when $x = \infty$, w and h are zero, so very large values of x would correspond to the origin in the Lie plane. We are looking for an integral curve, then, that goes from the origin in the Lie plane through the fourth quadrant to infinity. The only such curve is the separatrix S through the critical point P. Now in deriving (44) we calculated

$$x \frac{dw}{dx} = \frac{3}{18h - 16} (12w + 4h^2 - 18wh) \tag{42a}$$

and

$$x \frac{dh}{dx} = \frac{3}{18h - 16} (3w + 8h - 12h^2) \tag{42b}$$

From either of these equations we find that in the neighbourhood of P,

$$3 \frac{dx}{x} = \frac{dh}{h - \frac{1}{2}} \tag{43}$$

which means x must vanish as $(h - \frac{1}{2})^{1/3}$ at P, which is clearly impossible. So our effort to find an integral curve in the Lie plane fails.

The shock front is represented by the single point Q: (16/9, -64/27) and the profiles of Fig. 23 correspond to the part of the integral curve between Q and infinity. Numerical calculations indicate that Q does not lie on the separatrix S, a curve which seems to have little practical significance in this problem.

6.8 Transverse waves

In the commonest transverse wave problem, that of the stretched string, the string is pretensioned and the tension is considered to remain constant during the small transverse vibrations. If the string had not been pretensioned but had no slack initially, its tension could not have been considered constant, but would have depended on the amplitude of the transverse motion. This is because the transverse motion changes the length of the string. Such problems are more difficult to solve than simple stretched-string problems. But certain problems of this class exist whose solutions are given by similarity solutions; the problem of the shock-loaded elastic membrane mentioned in Section 1 of this chapter is the one dealt with by the author [DR70]. Figure 26 shows a schematic diagram of the setup. A completely flexible Hookean membrane in the form of a long ribbon is clamped along its two long edges O and P. Initially, the membrane has no slack but is under no pretension. At $t = 0$, the membrane is exposed suddenly to a uniform, steady

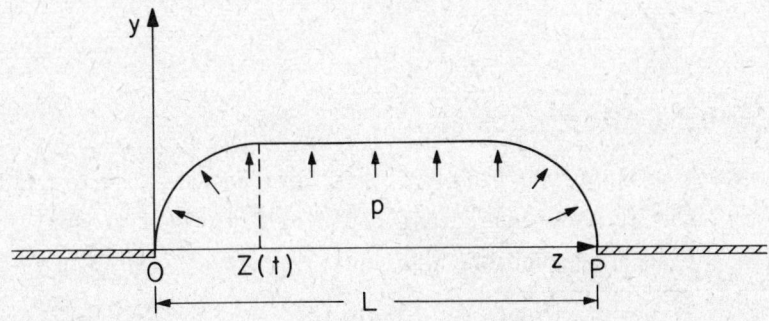

Fig. 26. Sketch of the shock-loaded membrane

pressure p. What is its subsequent motion?

The key to analyzing this problem is to realize that the velocity of transverse motion of the membrane elements is much less than their velocity of longitudinal motion. Consequently, the surface tension σ in the membrane is effectively uniform, having the instantaneous value

$$\sigma = \frac{Y}{L} \left[\int_0^L (1 + y_z^2)^{1/2} \, dz - L \right] \tag{44}$$

where Y is the elastic modulus of the membrane, L is its unstretched width, y is the transverse displacement, and z is the longitudinal coordinate (across the width). The pde of *small* transverse motions is, as usual,

$$\mu y_{tt} = p + \sigma y_{zz} \tag{45}$$

where μ is the mass per unit area of the membrane. If we expand (44) and keep only the leading term, (45) becomes

$$\mu y_{tt} = p + \frac{Y}{2L} y_{zz} \int_0^L y_z^2 \, dz \tag{46a}$$

or

$$y_{tt} = p + \frac{1}{2} y_{zz} \int_0^1 y_z^2 \, dz \tag{46b}$$

in special units in which $\mu = Y = L = 1$.

At time t, the centre part of the membrane $Z(t) < z < L - Z(t)$ has not yet been affected by the wave running in from the clamped edges.* So here the membrane is flat, $y_z = 0$, and $y = pt^2/2$. In the region $0 < z < Z(t)$, then,

$$y_{tt} = p + y_{zz} \int_0^{Z(t)} y_z^2 \, dz, \tag{47a}$$

the factor of 1/2 having disappeared from the last term because there are two disturbed regions, one at each edge, that contribute to the integral in (46b). It is convenient to introduce the auxiliary variable

---

* So the solution we are finding will describe only the early stage of the motion until the waves running inward from both clamped edges meet in the middle.

$$u = \frac{1}{2}pt^2 - y \tag{47b}$$

in terms of which (47a) becomes

$$u_{tt} = u_{zz} \int_0^{Z(t)} u_z^2 \, dz \tag{47c}$$

The equation of motion is then an integro-pde!

Equation (47c) is invariant to the group

$$\begin{aligned} u' &= \lambda^\alpha u \\ t' &= \lambda^\beta t \\ z' &= \lambda z \\ Z' &= \lambda Z \end{aligned} \tag{48a-d}$$

where α and β are subject to the linear constraint

$$2\alpha + 2\beta = 3 \tag{48e}$$

The actual values of α and β in this problem are determined by the boundary condition

$$u(0,t) = \frac{1}{2}pt^2 \tag{49}$$

which requires $\alpha = 2\beta$. Then, $\alpha = 1$ and $\beta = 1/2$. But before we use this information to try to find a similarity solution to (47c), let us again use the method of Section 4.15 to derive some useful information about $Z(t)$. If we add to (48) the additional transformation equation,

$$p' = \lambda^{\alpha-2\beta} p \tag{48f}$$

then (47c) and (49) are invariant to (48a-f). The extended group (48a-f) carries one problem with a pressure p into another of the same type with a pressure p'. Now, in general, for all such problems

$$Z(t) = F(t,p) \tag{50}$$

where F is an as yet undetermined function. But (50) must be invariant to (48a-f), from which it follows that

$$Z = Bp^{2/3}t^2 \tag{51}$$

where B is an as yet undetermined constant.

Now we proceed by setting

$$u = t^2 f\left(\frac{z}{t^2}\right) \tag{52}$$

When we substitute in (47c), we get

$$2\left(x^2 - \frac{1}{4}\int_0^X \dot{f}^2 dx\right)\ddot{f} - x\dot{f} + f = 0, \quad X \equiv Bp^{2/3} \tag{53}$$

The second term in parentheses is a constant independent of x, and its value can be found by the following argument. When $x = X \equiv Bp^{2/3}$, $f = 0$ since $x = X$ marks the wave front beyond which the membrane moves as if free. In order not to have a sharp crease in the membrane at $x = X$, we must also have $\dot{f}(X) = 0$. Now (53) is a second-order linear deq, so if $f = \dot{f} = 0$ at a regular point the solution must be identically zero. This means X must be a singular point of (53), and this fixes the value of the integral as

$$\frac{1}{4}\int_0^X \dot{f}^2 \, dx = X^2 \tag{54}$$

Then (53) becomes

$$2(x^2 - X^2)\ddot{f} - x\dot{f} + f = 0 \tag{55}$$

subject to the boundary conditions

$$f(X) = 0 \tag{56a}$$

$$\dot{f}(X) = 0 \tag{56b}$$

$$f(0) = p/2 \tag{56c}$$

The easiest way to solve (55) is to use the method of Section 3.1 based on (55)'s linearity and the fact that $f = x$ is a special solution. A straightforward but tedious calculation gives

$$f = \frac{p}{2\sqrt{X}}\left[(X^2 - x^2)^{1/4} - \frac{x}{2}\int_x^X (X^2 - x^2)^{-3/4} \, dx\right] \tag{57}$$

The reader may verify directly by substitution that (57) is a solution of (55) and that it does in fact satisfy all of (56a-c). If we now substitute (57) into (54) to determine X, or equivalently, B, we find (after two integrations by parts*)

$$B = \frac{1}{4}\left\{\int_0^1 ds \left[\int_s^1 (1-s'^2)^{-3/4} ds'\right]^2\right\}^{1/3} \tag{58a}$$

$$= \frac{1}{4}\left\{2\sqrt{\pi}\left[\frac{\Gamma(1/4)}{\Gamma(3/4)} - \sqrt{\pi}\right]\right\}^{1/3} = 0.4035 \tag{58b}$$

The condition $\dot{f}(X) = 0$, related to the membrane's having no crease at the moving front, has been added in an ad hoc way to the more or less natural boundary and initial conditions that we expect uniquely to determine the solution. If the other boundary conditions do in fact uniquely determine the solution, $\dot{f}(X) = 0$ should be a provable consequence of the ode and the other boundary and initial conditions. This is the case, and we prove it by noting that the quantity $\int_0^z u_z^2 dz = c^2$ *is not a function of* z. According to (48), c^2 transforms as

$$c'^2 = \lambda^{2\alpha-1} c^2 \tag{59}$$

Now c^2 must be a function only of p and t,

$$c^2 = G(p,t) \tag{60a}$$

and moreover the relation (60a) must be invariant to (48). Therefore the most general relation (60a) is

$$c^2 = A^2 p^{4/3} t^2 \tag{60b}$$

where A^2 is an as yet undetermined constant.

The quantity c is the wave propagation velocity, as we can easily see by introducing the auxiliary quantities $\xi = u_z$ and $\eta = u_t$ and rewriting (47c) as

---

* $\int_0^1 (1-s^2)^{-k} ds = \frac{\sqrt{\pi}}{2} \frac{\Gamma(1-k)}{\Gamma\left(\frac{3}{2}-k\right)}$

$$\eta_z = \xi_t \tag{61a}$$

$$\eta_t = c^2 \xi_z \tag{61b}$$

Multiplying (62a) by c and either adding or subtracting we obtain

$$(\eta - \int c\, d\xi)_t + c(\eta - \int c\, d\xi)_z = 0 \tag{61c}$$

$$(\eta + \int c\, d\xi)_t - c(\eta + \int c\, d\xi)_z = 0 \tag{61d}$$

From (61c,d) we can see that the positive and negative characteristics have the velocities $\pm c(= dz/dt)$, respectively. Using (60b), then, we see that the characteristics are given by

$$z = \pm \frac{A}{2} p^{2/3} t^2 + z_0 \tag{62}$$

where z_0 is the intercept of the characteristic on the z-axis.

A moment's thought (or a sketch of the wave diagram) will convince the reader that the positive characteristic through the origin, $z = (A/2)p^{2/3}t^2$, is the moving front dividing the disturbed from the as yet undisturbed part of the membrane. So $A/2 = B$, if we refer to (51). Then we find that (47c) can be written as

$$u_{tt} = 4B^2 p^{4/3} t^2 u_{zz} \tag{63a}$$

If we now substitute (52) into (63a) we find after a short calculation that

$$2(x^2 - B^2 p^{4/3})\ddot{f} - x\dot{f} + f = 0 \tag{63b}$$

which is the same result as (53) except that the integral has been given the value assigned to it in (54). Now we proceed as before, finding again the solution (57).

6.9 Elastic (Hookean) wire - transverse waves

With the machinery of calculation just established we can deal easily now with the transverse motion of a completely flexible Hookean wire mentioned at the beginning of Section 6.8. Figure 27 shows the conceptual setup. Again a weight is suspended from the wire, which initially is under no tension

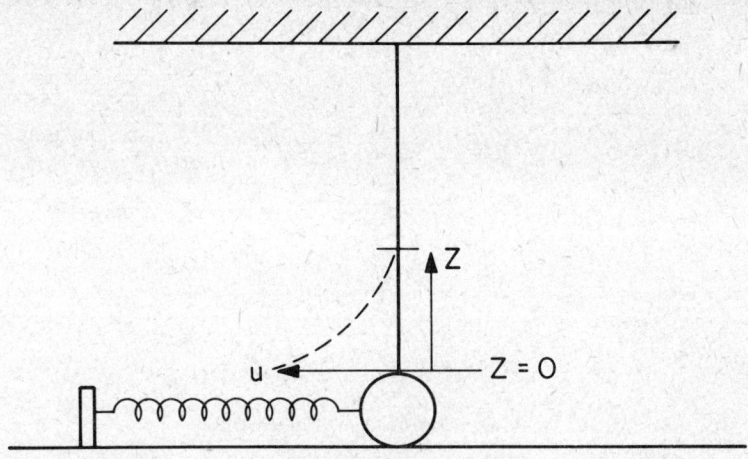

Fig. 27. Sketch of the spring accelerating the weight to the side

but has no slack. The weight is supported by a table and, starting at t = 0, it is uniformly accelerated sideways by the spring. Reasoning as before, we reach pde (47c) as the pde for the transverse displacement u.* The boundary condition (49) is now replaced by

$$u(0,t) = gt \tag{64}$$

where g is the uniform acceleration of the weight and (48f) is replaced by

$$g' = \lambda^{\alpha-\beta} g \tag{65}$$

Now (60b) becomes

$$c^2 = A^2 g^{4/3} t^{2/3} \tag{66}$$

so that (51) must be replaced by

$$Z = \frac{3}{4} A g^{2/3} t^{4/3} \tag{67}$$

---

* The special units are now those in which $\mu = L = 1$, $Y = 2$.

The boundary condition (64) requires that $\alpha = \beta = 3/4$. Then we must set

$$u = tf\left(\frac{z}{t^{4/3}}\right) \qquad (68)$$

After a short calculation we find

$$(x^2 - X^2)\ddot{f} + \frac{1}{4}x\dot{f} = 0 \qquad (69a)$$

where

$$X \equiv Z/t^{4/3} = \frac{3}{4}Ag^{2/3} \qquad (69b)$$

We must solve (69) subject to the boundary conditions

$$f(0) = g \qquad (70a)$$
$$f(X) = 0 \qquad (70b)$$

The deq (69) and boundary conditions can easily be integrated to give

$$f = C\int_x^X (x^2 - x^2)^{-1/8}\,dx \qquad (71)$$

as the reader can verify by direct substitution. The constant C is related to g and X by (70a):

$$C\int_0^X (x^2 - x^2)^{-1/8}\,dx = g \qquad (72)$$

The counterpart of (54) is

$$C^2\int_0^X (x^2 - x^2)^{-1/4}\,dx = \int_0^X \dot{f}^2 dx = A^2 g^{4/3} = \frac{16}{9}X^2 \qquad (73)$$

After some computation we find

$$A = 1.106 \qquad (74a)$$
$$C = 1.059\sqrt{g} \qquad (74b)$$

One interesting feature of the solution (71) is that $\dot{f}(X) = \infty$, which means that at the moving front the wire has a sharp, right-angled kink. Practically speaking, the kink is hardly discernible, and the profile (71) turns out to be close to linear in x.

6.10 Long waves in a channel

Another class of phenomena whose pdes give rise to characteristics is that of long waves in a channel. If u is the longitudinal flow velocity, h the local height of the liquid above the channel floor and z the longitudinal coordinate, then the equations of continuity and motion are, respectively,

$$h_t + (uh)_z = 0 \tag{75a}$$

$$u_t + uu_z + h_z = 0 \tag{75b}$$

in special units in which the fluid density ρ and the acceleration due to gravity g are taken as unity. If we multiply (75a) by c^{-1}, where $c^2 = h$, and add (75a) and (75b), we find the characteristic equation

$$\left(u_t \pm \frac{h_t}{\sqrt{h}}\right) + (u \pm \sqrt{h})\left(u_z \pm \frac{h_z}{\sqrt{h}}\right) = 0 \tag{76a}$$

which means

$$u \pm 2\sqrt{h} \quad \text{is conserved on} \quad dz = c_\pm \, dt \tag{76b}$$

and

$$c_\pm = u \pm \sqrt{h} \tag{76c}$$

The pdes (75) are invariant to the group

$$\begin{aligned} u' &= \lambda^\alpha u \\ h' &= \lambda^\gamma h \\ t' &= \lambda^\beta t \\ z' &= \lambda z \end{aligned} \tag{77a-d}$$

subject to the linear constraints

$$\alpha + \beta = 1 \tag{77e}$$

$$\gamma + 2\beta = 2 \tag{77f}$$

A problem in which a similarity solution occurs is the "breaking dam" problem. A semi-infinite channel is uniformly full of water to a height h_0. At $t = 0$, the restraining dam at $z = 0$ is removed. What is the subsequent motion of the water (see Fig. 28)?

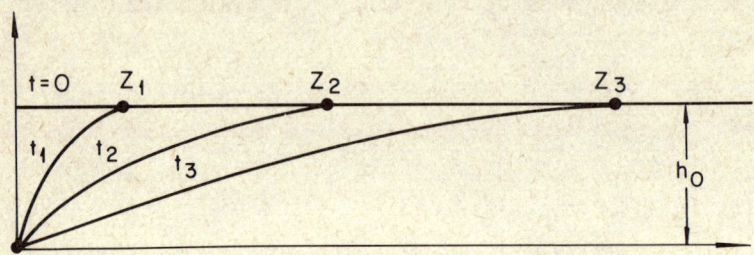

Fig. 28. Sketch of the water profiles behind the breaking dam at various times

The boundary condition $h(\infty,t) = h_0$ requires that γ be zero, so that then $\beta = 1$ and $\alpha = 0$. We therefore set

$$u = U\left(\frac{z}{t}\right) \tag{78a}$$

$$h = H\left(\frac{z}{t}\right) \tag{78b}$$

On account of the linear constraints (77e,f) we expect the odes for U and H to be invariant to the associated group

$$U' = \mu U \tag{79a}$$

$$H' = \mu^2 H \tag{79b}$$

$$x' = \mu x \quad (x \equiv z/t) \tag{79c}$$

From (79) it follows that the coordinate x_0 of the front separating the disturbed region from the undisturbed region (where $U = 0$ and $H = h_0$) must scale with $\sqrt{h_0}$, for x can depend only on h_0 and its dependence must be invariant to the associated group (79), which carries any similarity solution into another one with a different initial height. In fact, it follows from (76c) that $x_0 = \sqrt{h_0}$.

The odes for U and H are

96

$$(U - x)\dot{H} + H\dot{U} = 0 \tag{80a}$$

$$\dot{H} + (U-x)\dot{U} = 0 \tag{80b}$$

Now these equations are homogeneous in $\dot{U}$ and $\dot{H}$ so, if they are to have a nontrivial solution, the determinant of the coefficients must vanish:

$$(U - x)^2 - H = 0 \tag{81}$$

From (81), too, it follows at once that $x_0 = \sqrt{h_0}$, so that even if we had not found this result in another way above, it would have fallen out at this point. When (81) is fulfilled, both odes (80) are the same. It is easy to solve them and find

$$U = 2(x - \sqrt{h_0})/3 \tag{82a}$$

$$H = (x + 2\sqrt{h_0})^2/9 \tag{82b}$$

6.11 Travelling waves

Suppose a second-order pde in t and z is invariant to the one-parameter family of translations

$$\begin{aligned} t' &= t + \lambda, \quad -\infty < \lambda < \infty \\ z' &= z + \alpha\lambda, \quad 0 < \alpha < \infty \end{aligned} \tag{83}$$

The most general function of z and t invariant to (83) is $f(z - \alpha t)$, where f is as yet undetermined. Such a solution is called a travelling wave solution. If we substitute this function into the pde we get an ode for f involving the parameter α. Usually the boundary conditions lead to an eigenvalue problem for the acceptable values of α. Let one such value be α_0. Then the solution of the ode will have the form $f(z - \alpha_0 t)$. If we transform this with the group (83) for a value of $\alpha \neq \alpha_0$, we must get another solution of the pde, namely,

$$f[z + \alpha\lambda - \alpha_0(t + \lambda)] = f[z - \alpha_0 t + (\alpha - \alpha_0)\lambda] \tag{84}$$

Thus if $f(x)$ is a travelling wave solution so is $f(x + \mu)$, where $\mu \equiv (\alpha-\alpha_0)\lambda$. But this means that the ode for $f(x)$ is invariant to the associated group

$$f' = f$$
$$\qquad -\infty < \mu < \infty \qquad (85)$$
$$x' = x + \mu$$

An invariant and first differential invariant of (85) are f and $\dot{f}$, respectively. The ode for f can then be written as the *first-order* ode

$$\frac{d\dot{f}}{df} = F(f,\dot{f}) \qquad (86)$$

where F is a function that depends on the form of the pde. The first-order ode, like all first-order odes, can be studied through its Lie plane.

One application in which the travelling wave is of practical importance is that of a quenching superconductor. A non-superconducting (high-temperature) zone, once established in a superconducting wire, spreads as a travelling wave. The velocity of the wave front is an important datum in understanding heat transfer from the wire to its surroundings and also in the design of protection circuits for magnets wound with superconducting wire.

The pde that governs this application is

$$c_t = c_{zz} + Q(c) \qquad (87)$$

where $Q(c)$ has the form shown in Fig. (29a). Figure (29b) shows the form

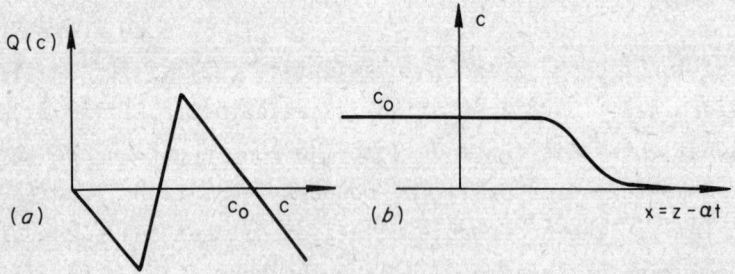

Fig. 29. Sketches showing the forms of (a) the function $Q(c)$ and (b) the travelling wave solution

we seek for the travelling wave solution. Equation (86) has the particular form

$$\frac{d\dot{f}}{df} = -\alpha - \frac{Q(f)}{\dot{f}} \tag{88}$$

Figure 30 shows the Lie plane of the associated ode (88). There are three critical points, 0, Q and P, all on the f-axis and located at the three roots of the function Q(f). The singularities 0 and P are saddle points and the

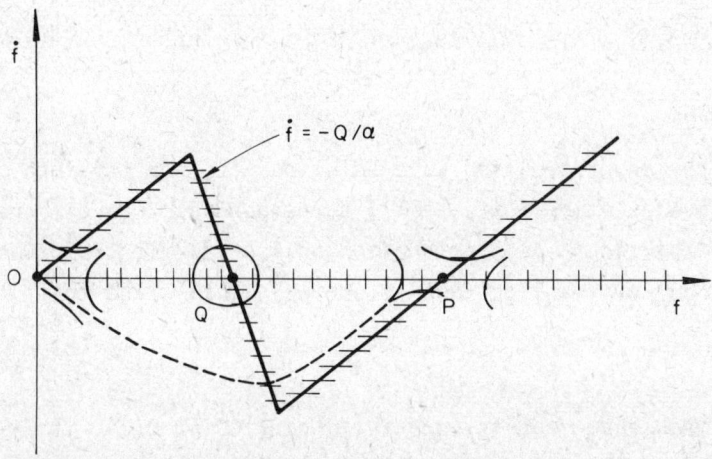

Fig. 30. The Lie plane of ode (88)

singularity Q is a vortex point. A solution of the type shown in Fig. 29b must correspond to an integral curve in Fig. 30 that begins at the origin and ends on the f-axis. The only curve that can do this is a separatrix going from 0 to P, as shown in Fig. 30. It turns out that the two separatrices from 0 and from P only join for a certain value of α, although this is by no means clear from the analysis carried out so far. (The author has carried out the solution of the associated ode (88) for a particular form of Q(c) quantitatively similar to Fig. 29b and of practical importance.) But one conclusion that has already been proved is that the value of c far to the left of the travelling front, c_0, is the upper root c_0 of Q(c).

Another useful conclusion that can be proved about the travelling wave solutions of (87) concerns the dependence of α on the amplitude of Q(c). Suppose we replace Q(c) by aQ(c), where $0 < a < \infty$ is a multiplier that changes the amplitude of a. How does α depend on a? Equation (87) is invariant not only to the translations (83) but also to the stretching transformations

$$c' = c$$
$$t' = \lambda^2 t$$
$$z' = \lambda z \qquad \text{(89a-d)}$$
$$a' = \lambda^{-2} a$$

Equations (89a-d) carry any solution of (87) into another solution corresponding to another value of a. If we add to (89a-d) the further equation

$$\alpha' = \lambda^{-1} \alpha \qquad \text{(89e)}$$

then (89a-e) will carry any travelling wave solution $f(z' - \alpha't')$ into another travelling wave solution $f[\lambda(z - \alpha t)]$ corresponding to a different α. Now for all such solutions, α is a function only of a. Its functional dependence on a must be invariant to (89a-e), so we quickly find that

$$\alpha = \text{constant} \times \sqrt{a} \qquad \text{(90)}$$

From (90) it is clear that if α is zero for a particular $Q(c)$, it will remain zero even if the amplitude of $Q(c)$ changes. So the vanishing of α depends only on the shape of $Q(c)$. From (88) with $\alpha = 0$ it follows that

$$\int_0^{c_0} Q(f) df = 0 \qquad \text{(91)}$$

i.e., the areas of the two lobes of $Q(c)$ must be equal when α vanishes. This is the famous equal-area theorem of Maddock, James and Norris [MA69].

7 Miscellaneous topics

7.1 Approximate solutions: diffusion in cylindrical geometry

In Section 5.4, we saw an attempt to apply the methods used for obtaining similarity solutions in a problem to which they should not strictly apply. The failure there of the method of local similarity might tempt us to think that similarity methods are of no help in problems in which all the conditions of applicability are not met. But a deeper look shows that similarity methods can often be used to get valuable information about asymptotic behaviour in such problems. An excellent illustration is the clamped-flux problem in cylindrical geometry.

Suppose that at $t = 0$ the cylindrical surface $r = R$ suddenly begins emitting a steady heat flux Q. What is the resulting temperature at the cylindrical surface? The pde describing the problem is

$$T_t = \frac{1}{r}(rT_r)_r \tag{1}$$

in special units in which the thermal conductivity, density and specific heat of the material are taken to be unity. The associated boundary conditions are

$$T_r(R,t) = -Q, \quad t > 0 \tag{2a}$$

$$T(r,0) = 0, \quad r > R \tag{2b}$$

$$T(\infty,t) = 0, \quad t > 0 \tag{2c}$$

The pde is invariant to the group

$$T' = \lambda^\alpha T \tag{3a}$$

$$t' = \lambda^2 t \tag{3b}$$

$$r' = \lambda r \tag{3c}$$

but the boundary conditions are not because of the appearance of the value $r = R$.

If we add to (3a-c) the transformation equations

$$R' = \lambda R \qquad (3d)$$

$$Q' = \lambda^{\alpha-1} Q \qquad (3e)$$

then equations (3a-e) transform (1) and (2) from the description of a problem with a flux Q at radius R to the description of a problem with a flux Q' at radius R'. Now the temperature of the cylindrical surface can only be a function of t, R, and Q, and the functional relationship must be the same for all problems of this class. Hence it must be invariant to (3a-c). It is easy to see, then, that the most general relation possible is

$$T(R,t) = Q \sqrt{t}\, F\left(\frac{t}{R^2}\right) \qquad (4)$$

where F is an as yet undetermined function. We shall not be able to determine F exactly, but we shall be able to determine its asymptopic behaviour for both large and small t/R^2 and thereby to estimate it for all values with accuracy sufficient for practical purposes.

Very early, the heat from the source surface $r = R$ has not diffused very far. When the thickness of the heated layer is still $\ll R$, the curvature of the heated surface should not matter. So the problem goes over, for short times, to its plane analogue

$$T_t = T_{zz} \qquad (5a)$$

$$T_z(0,t) = -Q, \quad t > 0 \qquad (5b)$$

$$T(z,0) = 0, \quad z > 0 \qquad (5c)$$

$$T(\infty,t) = 0, \quad t > 0 \qquad (5d)$$

where $z = r - R$. We solved this problem in Section 3.1; using the results from there we find $F = 2/\sqrt{\pi}$ for small t/R^2.

For long times we proceed as follows. The "troublesome" boundary condition (2a) is equivalent to the condition

$$\frac{d}{dt} \int_R^\infty Tr \, dr = RQ \tag{6a}$$

or

$$\int_R^\infty Tr \, dr = RQt \tag{6b}$$

as we can see by integration of the pde (1) over the region $R < r < \infty$. Late in the history of the problem, when the temperature profile has spread very far from $r = R$, the value of the integral in (6b) should be little affected if we extend the lower limit to zero. If we do so we again have a totally invariant problem. Then equation (6b) requires $\alpha = 0$, so we take

$$T = G\left(\frac{r}{t^{1/2}}\right) \tag{7}$$

and obtain

$$\ddot{G} + \left(\frac{1}{x} + \frac{x}{2}\right)\dot{G} = 0, \quad x = \frac{r}{\sqrt{t}} \tag{8a}$$

$$G(\infty) = 0 \tag{8b}$$

$$\int_0^\infty G \, x \, dx = RQ \tag{8c}$$

Equation (8a) integrates at once to give

$$\dot{G} = \text{const.} \times \frac{e^{-x^2/4}}{x} \tag{9}$$

Integrating by parts turns (8c) into

$$\int_0^\infty \dot{G} \, x^2 \, dx = -2RQ \tag{10}$$

from which it follows that the constant in (9) equals $-RQ$. Finally, then

$$G(x) = RQ \int_x^\infty \frac{e^{-x^2/4}}{x} \, dx = \frac{RQ}{2} E_1\left(\frac{x^2}{4}\right) \tag{11}$$

where E_1 is the exponential integral discussed by Abramowitz and Stegun [AB68]. By combining (11) with our previous result we get

$$F(\tau) = \frac{2}{\sqrt{\pi}} \qquad \tau \ll 1$$
$$= \frac{E_1(1/4\tau)}{2\sqrt{\tau}} \qquad \tau \gg 1 \qquad \Bigg\} \quad \tau = \frac{t}{R^2} \qquad (12)$$

Figure 31 shows $T(R,t)/QR$ as a function of t/R^2. The solid lines are the short- and long-time asymptotes calculated from the values of F given in (12). The dashed curve is a graphical interpolation between the two asymptotes.

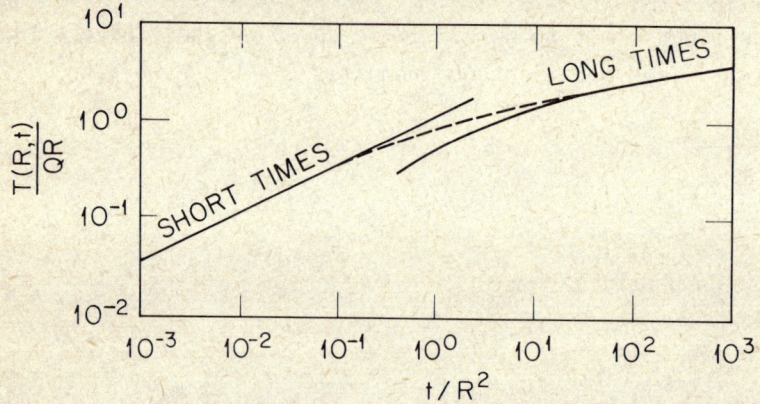

Fig. 31. Wall temperature versus time for constant wall flux in cylindrical geometry

It is no rare thing for the limiting behaviour of a physical system to be describable in terms of a similarity solution even though the general behaviour is not. For example, in laminar, forced convection heat transfer in a pipe, the well-known Levich-Levêque [LE62] solution for the entrance region is a similarity solution valid when the thermal boundary layer is small compared with the pipe radius. The concentration polarization solution of Section 3.3 is a quite similar entrance-region solution. The solution (6.57) of the shocked membrane problem is an early-time asymptote. Barenblatt [BA79] has emphasized the utility of such asymptotic similarity solutions and discussed them at length in his book.

7.2 Diffusion in cylindrical geometry (cont'd): clamped temperature

If we clamp the temperature rather than the flux, we have the same pde (1) and keep the boundary conditions (2b) and (2c), but the boundary condition (2a) is replaced by

$$T(R,t) = T_0 \quad t > 0 \tag{13}$$

If we add to (3a-c) the transformation equations

$$T_0' = \lambda^\alpha T_0 \tag{14a}$$

$$R' = \lambda R \tag{14b}$$

then equations (3a-c) and (14a,b) transform (1), (2b,c) and (13) from the description of a problem with a temperature T_0 at radius R to one with temperature T_0' at radius R'. Then, by a now familiar argument, the surface flux $Q = -T_r(R,t)$ must be given by

$$Q = T_0 t^{-1/2} F\left(\frac{t}{R^2}\right) \tag{15}$$

where again F is an as yet undetermined function.

As before, the curvature of the heated surface can be ignored for very short times. We solved the plane variation of the clamped-temperature problem in Section 3.1 also. Using those results we find $F = \pi^{-1/2}$ for short times.

In the clamped-flux problem, the "troublesome" boundary condition (2a) could be replaced for long times by another boundary condition that allowed a similarity solution. But this does not happen with the "troublesome" clamped-temperature boundary condition (13). However, in this problem we can use successfully the method of local similarity, which proved so disappointing in Section 5.4. We proceed by setting

$$T = H(t)\, G\left(\frac{r}{t^{1/2}}\right) \tag{16}$$

Substituting (16) into (1), we find the principal ode

$$\ddot{G} + \left(\frac{1}{x} + \frac{x}{2}\right)\dot{G} = \frac{\dot{H} t}{H} G, \quad x = \frac{r}{t^{1/2}} \tag{17}$$

If H varies slowly enough with t, the right-hand side of (17) may be neglected. Then $G = E_1(x^2/4)$ (any constant of integration may be subsumed in H). If $T(R,t)$ is to equal T_0, then $H(t)$ must equal $T_0/E_1(R^2/4t)$, so that

$$T = T_0 \frac{E_1(r^2/4t)}{E_1(R^2/4t)} \tag{18}$$

Then

$$Q = -T_r(R,t) = \frac{T_0}{R} \frac{2e^{-R^2/4t}}{E_1(R^2/4t)} \tag{19a}$$

so that

$$F(\tau) = \frac{2\sqrt{\tau}\, e^{-1/4\tau}}{E_1(1/4\tau)} \quad \text{for large } \tau = t/R^2 \tag{19b}$$

Now we must verify that $G\dot{H}/H$ is less than the other two (equal) terms in Eq. (17). A short calculation shows that

$$\left| \frac{G}{\ddot{G}} \cdot \frac{\dot{H}t}{H} \right| = \frac{E_1(x^2/4)}{e^{-x^2/4}(1 + 2/x^2)} \cdot \frac{e^{-R^2/4t}}{E_1(R^2/4t)} \tag{20}$$

Since $e^{x^2/4} E_1(x^2/4)$ is a monotone decreasing function of x, its largest value occurs when x is smallest, i.e., when $r = R$. Thus

$$\left| \frac{G}{\ddot{G}} \cdot \frac{\dot{H}t}{H} \right| < \frac{R^2}{8t + R^2} < \frac{R^2}{8t} \tag{21}$$

Thus for $t/R^2 \gg 1$, i.e. for long enough time, the right-hand side of (17) may be neglected. Figure 32 shows the heat flux at $r = R$ as a function of t/R^2. Again the solid lines are the asymptotic similarity solutions and the dashed curve is a graphical interpolation. Shown for comparison are exact results for the slab and sphere. (The sphere problem can be converted to the slab problem by using rT as a new variable.)

The above analysis is equally valid when T_0 is a sufficiently slowly varying function of t. How slowly can be determined by noting that if T_0 is time dependent, the term $\dot{T}_0 t/T_0$ is added to the second factor on the right in (20). Eventually, for large t, this additional term leads to a second term on the right-hand side of (21), namely,

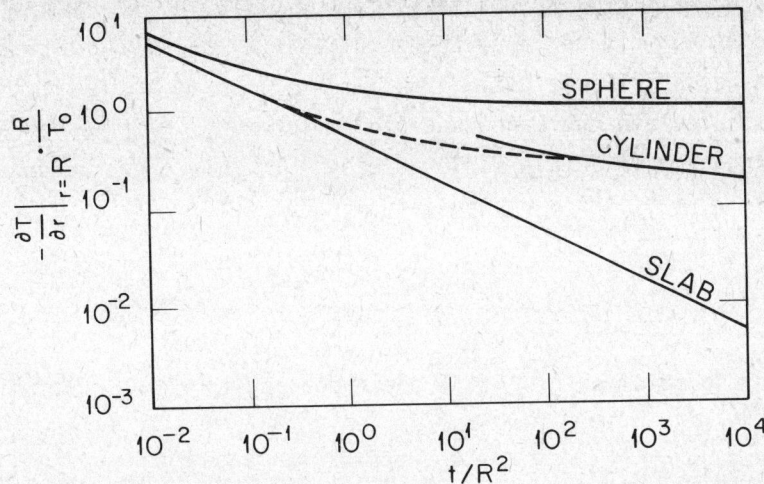

Fig. 32. Wall flux versus time for constant wall temperature in cylindrical geometry

$$\frac{R^2 \dot{T}_0}{2 T_0} \cdot \ln\left(\frac{4t}{R^2}\right) \tag{22}$$

which must also be small. It is clear from (22) then that the method of local similarity is valid for a wide variety of boundary values $T_0(t)$.

7.3 The method of local similarity

The factorized (variables separated) form (16) works well in linear problems, but we must return to a more general form to treat nonlinear problems. This is best illustrated by an example. Consider the nonlinear clamped-flux problem solved in Section 4.2, but suppose now the flux $-(CC_z)_{z=0}$ is clamped not at a constant but at some function of time $b(t)$. Following the idea of Sparrow et al. [SP70], we set

$$C = t^{1/3} y(x,t), \quad x = z/t^{2/3} \tag{23}$$

Now we obtain for y the *partial* differential equation

$$3(yy_x)_x = y - 2xy_x + 3ty_t \tag{24}$$

If we ignore the last term, we get the *ordinary* differential equation (4.11) for y. Accordingly, we take for y the solution y(x) of (4.11) corresponding to the instantaneous value of b(t).

Now we must calculate the magnitude of the term $3ty_t$. If we had an explicit solution for y(x) this would be easy. In this case, we do not, but the invariance of (4.11) to the associated group (4.13) will be all we need. According to (4.13)

$$y(x) = \mu^2 \, y_0\left(\frac{x}{\mu}\right) \tag{25a}$$

where $y_0(x)$ is the solution of (4.11) corresponding to b(0). Thus since $y(0) = C^{-2/3} b^{2/3}$ (Eq. 4.18),

$$\mu = \left[\frac{b(t)}{b(0)}\right]^{1/3} \tag{25b}$$

Differentiating (25a) partially with respect to t, we find

$$y_t = \mu \dot{\mu}(2y_0 - x\dot{y}_0)\big|_{x/\mu} \tag{26a}$$

so that, using the value μ in (25b), we find

$$\frac{3ty_t}{y} = \frac{\dot{b}t}{b}\left(2 - \frac{x\dot{y}_0}{y_0}\right)\bigg|_{x/\mu} \tag{26b}$$

or

$$\frac{3ty_t}{(-2xy_x)} = \frac{\dot{b}t}{b}\left(-\frac{y_0}{x\dot{y}_0} + \frac{1}{2}\right)\bigg|_{x/\mu} \tag{26c}$$

Thus if $\dot{b}t/b < 1$, we can use (26) to show that near $x = 0$, $3ty_t \ll y$, while near the first root of y_0, $3ty_t \ll (-2xy_x)$. Thus, when $\dot{b}t/b \ll 1$ the term $3ty_t$ is always smaller than one or the other of its companion terms on the right-hand side of (24) and can be neglected. Then,

$$C(0,t) = C^{-2/3} t^{1/3} b^{2/3}(t), \quad \dot{b}t/b \ll 1 \tag{27}$$

An interesting by-product of the condition $\dot{b}t/b \ll 1$ is that the expression (27) is always monotone increasing. For,

$$C_t(0,t) = \frac{2}{3} C^{-2/3} (b/t)^{2/3} \left[\frac{1}{2} + \frac{\dot{b}t}{b}\right] \tag{28}$$

so $C_t(0,t)$ can never vanish if $\dot{b}t/b \ll 1$.

We should be able to improve on the estimate (27) if we use a more flexible trial function. A convenient form is $p(t) y_1[q(t)x]$ where $y_1(x)$ is the solution of (4.11) corresponding to the instantaneous value of $b(t)$ and $p(t)$ and $q(t)$ are as yet undetermined functions of t.

The boundary condition at $z = 0$ puts one constraint on p and q:

$$-b(t) = (CC_z)_{z=0} = p^2 q\, y_1(0)\dot{y}_1(0) = -p^2 q\, b(t) \tag{29a}$$

the last equality following because y_1 is the similarity solution corresponding to the instantaneous value of $b(t)$. Thus

$$p^2 q = 1 \tag{29b}$$

To determine p and q we need one more relationship, which we get using the integral method. If we integrate (24) from $x = 0$ to $x = \infty$ we find, after an integration by parts,

$$b(t) = \int_0^\infty y\,dx + t\,\frac{d}{dt}\int_0^\infty y\,dx \tag{30a}$$

from which it follows easily that

$$\int_0^\infty y\,dx = \frac{1}{t}\int_0^t b(t')\,dt' \tag{30b}$$

If we substitute $y = py_1(qx)$ for y, we obtain

$$\frac{p}{q}\int_0^\infty y_1(x')\,dx' = \frac{1}{t}\int_0^t b(t')\,dt' \tag{31}$$

Now $\int_0^\infty y_1(x')\,dx' = b(t)$ as we can see, for example, by integrating (4.11) from $x = 0$ to $x = \infty$. Then

$$p^3 = \frac{p}{q} = \frac{\int_0^t b(t')\,dt'}{t\,b(t)} \tag{32}$$

This leads to the improved estimate

$$C(0,t) = C^{-2/3} t^{1/3} b^{2/3}(t) \left[\frac{\int_0^t b(t')\,dt'}{t\,b(t)}\right]^{1/3} \tag{33}$$

We can check the degree of improvement of (33) over (27) by taking $b(t) = t$. For this choice, a similarity solution of the form $C = t y_2(z/t)$ exists. The computation of y_2 follows closely the method outlined in Section 4.2; numerical integration of the associated deq along the separatrix gives $y_2(0) = 1.000$. According to (27), $C(0,t) = 1.294\, t$, whereas according to (33), $C(0,t) = 1.027\, t$. The error in (27) then is 29% but the error in (33) is only 2.7%. So the added flexibility of the trial form has enabled us to reduce the error in our estimate of $C(0,t)$ by about a factor of 10.

Extending the flexibility of the trial function in this way does not always bring an improvement in accuracy. Consider the problem of the boundary layer on a flat plate with uniform suction or injection (Section 5.4). Let us try

$$f(\xi,\eta) = p(\xi)\, F[q(\xi)\eta] \tag{34a}$$

where

$$2\dddot{F} + F\ddot{F} = 0 \tag{34b}$$

$$\dot{F}(0) = 0 \tag{34c}$$

$$\dot{F}(\infty) = 1 \tag{34d}$$

$$F(0) = h(\xi) \tag{34e}$$

The boundary conditions (5.34) lead to the following connections between p, q and h:

$$pq = 1 \tag{35a}$$

$$h\frac{d}{d\xi}(\xi p) + 2\xi = 0 \tag{35b}$$

Substituting (34a) into (5.33) we find

$$1 - q^2 + \xi q \dot{p} = 0 \tag{35c}$$

From these equations it follows that, most generally,

$$1 - p^2 = \text{const}/\xi^2 \tag{36}$$

Since we want $p(0) = 1$, we must choose the constant of integration to be zero, and then $p = q = 1$ and $h = -2\xi$. Thus the only ansatz of the form (34a) that will work is the one already used in Section 5.4.

7.4 Eigenvalue problems

Perhaps the best-known eigenvalue problem in the literature of mathematical physics is the one-electron spectrum. For s-waves,

$$\frac{1}{r^2} \frac{d}{dr}\left(r^2 \frac{dc}{dr}\right) - \frac{V}{r} c = Ec \qquad (37a)$$

$$c(0) = 0, \quad c(\infty) = 0 \qquad (37b)$$

Since the eigenvalues do not depend explicitly on r and c, they are at most functions of the strength V of the interaction. Now (37) is invariant to the group

$$\begin{aligned} c' &= \lambda^\alpha c \\ r' &= \lambda r \\ V' &= \lambda^{-1} V \\ E' &= \lambda^{-2} E \end{aligned} \qquad (38)$$

The functional relationship of E and V must be invariant to (38), so that, most generally, $E = \text{constant} \cdot V^2$. The key point here, as in Section 4.15, is the dependence of the eigenvalue on sufficiently *few* parameters that invariance to the principal group serves entirely to determine the functional dependence.

A more complicated nonlinear eigenvalue problem is one related to the quenching superconductor studied in Section 6.11. The superconductor there was supposed to be in contact with liquid helium which gave $Q(c)$ the shape it had in Fig. (29a). If the superconductor is potted in plastic, also a case of technological importance, then $Q(c)$ has the shape shown in Fig. 33.

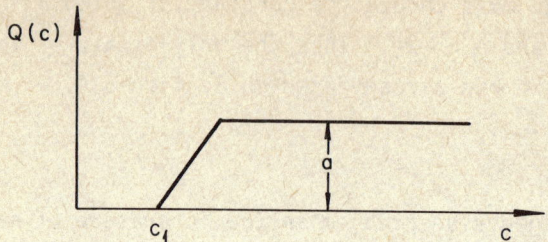

Fig. 33. Sketch showing the form of $Q(c)$ for a potted superconductor

Suppose at $t = 0$ we put an instantaneous pulse of heat q in the plane $z = 0$, i.e., suppose we choose the initial distribution of c to be

$$c = \frac{q \exp(-z^2/4t)}{(4\pi t)^{1/2}} \quad \text{(small t)} \tag{39}$$

If q is small enough, diffusion, which dominates the behaviour of (6.87) at small times when there are large gradients, will reduce the central value of c below c_1 so that $Q(c)$ will be zero everywhere. From that time on, c will obey the ordinary diffusion equation. If q is large enough, diffusion will never dominate the source term and $Q(c)$ will never vanish everywhere. These two qualitatively different regimes are separated by a limiting value of q, which is of technological significance. How does it depend on the amplitude of $Q(c)$ (a quantity under the control of the superconductor designer)?

Equations (6.87) and (39) are invariant to the group of transformations

$$\begin{aligned} c' &= c \\ z' &= \lambda z \\ t' &= \lambda^2 t \\ q' &= \lambda q \\ a' &= \lambda^{-2} a \end{aligned} \tag{40}$$

Since the limiting value of q depends only on a (for a fixed form of $Q(c)$) we have at once

$$q_{lim} = \text{const} \times a^{-1/2} \tag{41}$$

Problems

Chapter 2

1. Find a stretching group to which $x\dot{y} = yf(xy)$ is invariant, use Lie's formula for the integrating factor, and integrate.

2. Find a translation group to which the Ricatti equation $\dot{y} = (x + y)^2$ is invariant, rework the reasoning of Section 2.2 to find an integrating factor, and solve.

3. Sketch the direction field and integral curves of the equation $(\ln y + 2x - 1)\dot{y} = 2y$. Does it have a separatrix? (HINT: Find a mixed stretching-translation group leaving the deq invariant and repeat the reasoning of Section 2.5.)

4. Find a stretching group to which the Ricatti equation $x^2(\dot{y} - y^2) = 1/4$ is invariant and use the method of Section 2.6 to separate variables and integrate.

5. Find a mixed translation-stretching group to which the Poisson-Boltzmann equation in cylindrical coordinates $\ddot{y} + \frac{\dot{y}}{x} = e^y$ is invariant. Construct a differential invariant and a first differential invariant and use them to reduce the order of the deq. Solve the resulting first-order deq and select the integral curve that corresponds to solutions $y = y(x)$ that are regular at the origin. Find an integrating factor, integrate again, and find an explicit formula for these regular solutions.

Chapter 3

6. Solve the linear diffusion equation in a infinite medium with a continuous source located in the plane $z = 0$ whose integrated strength is proportional to the elapsed time. What is the time dependence of the temperature at the location of the source?

7. The linear partial differential equation $C_t = (\sqrt{z}\, C_z)_z$ occurs in the calculation of diffusion to a solid particle freely falling in a viscous

liquid; the boundary conditions are $C(0,t) = 0$, $C(\infty,t) = 1$, $C(z,0) = 1$. The diffusional flux into the particle is proportional to $t^{1/3} \sqrt{z}\, C_z(z,t)|_{z=0}$. Find this quantity by an analysis similar to that of Section 3.3.

Chapter 4

8. The pde $C_t = (C^m)_{zz}$ arises in the calculation of the current distribution in a superconductor suddenly charged with current. Find the similarity solution describing the infinite-medium problem of an instantaneous heat pulse in the plane $z = 0$ at time $t = 0$.

9. Integrate eq. (4.15) and show explicitly that $v \sim \sqrt{u}$ for large u. (HINT: Consider the three possibilities in the footnote on p. 30.)

10. The ordinary diffusion equation (3.1) is invariant to the one-parameter family of stretching groups (3.2). Use the reasoning of Section 4.3 to determine the associated group for eq. (3.9). Introduce an invariant and a first differential invariant and obtain the associated ode in the case $\alpha = 0$. Is it easier to solve than the linear ode (3.11a)? Try using the invariant and first differential invariant $u = x$, $v = (\dot{y}/y)e^{x^2/4}$. Can you solve the resulting associated ode and eventually obtain eq. (3.12)?

11. Carry out the computations leading to eq. (4.28).

12. Find a family of stretching groups that leaves the pde $p_t = -(\sqrt{-p_z})_z$ invariant and show for any similarity solution that $V \equiv \sqrt{-p_z}|_{z=0}$ is proportional to $p^{2/3}(0,t)\, t^{-1/3}$. (HINT: Use the associated group.)

13. Consider the clamped-temperature problem for the pde $C_t = (e^C C_z)_z$; the boundary conditions are $C(0,t) = C_0$, $C(\infty,t) = C_\infty$, and $C(z,0) = C_\infty$. Find a family of groups that leaves the pde invariant. Repeat the reasoning of Section 4.3 and show that an associated group exists. If y is the dependent variable in the principal ode, show from the associated group that $y(0) - y(\infty)$ is a function of $\dot{y}^2(0)e^{y(0)}$.

Chapter 5

14. When the boundary-layer flow is over a curved surface, the term UU_x must be added to the right-hand side of eq. (5.1a), where $U(x)$ is the potential flow far from the curved surface. For what functional dependences of U on x do similarity solutions exist? Does an associated group exist for any of

these cases?

15. Write down the appropriate similarity solution for the clamped-flux case of the thermal boundary-layer problem of Section 5.5. Find the principal ode for the temperature. What are the boundary conditions? Is the problem a two-point boundary-value problem? Use the linearity of the principal deq and the fact that the Blasius function $f \sim \eta - 1.7208$ for very large η to find the asymptotic form of the temperature and circumvent the two-point difficulty.

Chapter 6

16. Show that if the coupled pair of first-order odes $\dot{y} = Y(x,y,z)$, $\dot{z} = Z(x,y,z)$ is invariant to the group of transformations $y' = \lambda^\alpha y$, $z' = \lambda^\beta z$, $x' = \lambda x$, then introducing the invariants $u = y/x^\alpha$ and $v = z/x^\beta$ as new variables will reduce the pair of odes to a single, first-order ode for v in terms of u. (HINT: Proceed as we did in Section 2.7.)

17. Suppose the positive characteristics in the wave diagram of Fig. 20 all emanated fan-like from the origin O. What solution for v and η in terms of a and t would this diagram correspond to? Compare this solution with the exceptional solution corresponding to point P in Fig. 18.

18. What variations of σ with η allow eqs (6.3) and (6.4b) to be invariant to a stretching group? Show that such a stretching group automatically leaves the shock conditions (6.31) invariant. What assumption did you have to make about how the shock velocity transforms under the stretching transformations?

19. Suppose the surface tension in the membrane of Section 6.8 varies as the mth power of the longitudinal strain. How does the time for the waves running in from the edges to reach the centre scale with applied pressure?

Chapter 7

20. Repeat the analysis of Section 7.3 for the clamped-flux problem in superfluid helium (Section 4.9) and obtain a formula for the wall temperature when the imposed flux varies slowly with time.

Solutions to problems

1. The group is $x' = \lambda x$, $y' = \lambda^{-1} y$. According to Lie's theorem, Section 2.2, $[xy(f(x,y) + 1)]^{-1}$ is an integrating factor. The deq can then be integrated to give $\ln y = \int_C^{xy} \frac{f(u) \, du}{u(f(u) + 1)}$ where C is an arbitrary constant.

2. The group is $x' = x + \lambda$, $y' = y - \lambda$. An integrating factor is $(M - N)^{-1}$, which in the problem at hand turns out to be $[(y + x)^2 + 1]^{-1}$. Now we integrate to find $y = \tan(x + C) - x$, where C is a constant of integration.

3. The mixed stretching-translation group is $x' = x - \frac{1}{2} \ln \lambda$, $y' = \lambda y$. Invariant curves obey the identity $g(x - \frac{1}{2} \ln \lambda, \lambda y) = 0$ for any λ. From this it follows that $dy/dx = -g_x/g_y = -2y$, so that $2y(\ln y + 2x) = 0$ becomes the counterpart of eq. (2.22). Thus the invariant curves are $y = 0$ and $y = e^{-2x}$; the latter is the separatrix.

4. The group is $x' = \lambda x$, $y' = \lambda^{-1} y$, and a group invariant is $u = xy$. Using it as a new dependent variable, we find $dx/x = du/(u + \frac{1}{2})^2$, which is separable. Integrating, we find $y = -\frac{1}{x}(\frac{1}{2} + \frac{1}{\ln(x/C)})$, where C is a constant of integration.

5. The group is $x' = \lambda x$, $y' = y - 2 \ln \lambda$; the invariant and first differential invariant are $u = x^2 e^y$ and $v = x\dot{y}$. The associated deq is $dv/du = (v + 2)^{-1}$ which can be integrated to give $v^2 + 4v = 2u$; the constant of integration has been chosen to make $v = 0$ when $u = 0$ because y and $\dot{y}$ are supposed to be regular at $x = 0$. We can solve for v and find $v = (4 + 2u)^{1/2} - 2$ (the positive sign of the square root is taken so that $v = 0$ when $u = 0$). This can be written $x\dot{y} + 2 - (4 + 2x^2 e^y)^{1/2} = 0$ which is also invariant to the group cited above. Thus $[x(4 + 2x^2 e^y)^{1/2}]^{-1}$ is an integrating factor. A lengthy calculation now shows that

$$y = \ln\left[(\frac{x^2 + b^2}{x^2 - b^2})^2 - 1\right] - \ln x^2 + \ln 2.$$

6. The source condition is $\int_{-\infty}^{+\infty} C\,dz = t$ so $\alpha = 1$ and we can use eq. (3.15) for the solution. The source condition in terms of y is $\int_0^\infty y\,dx = \frac{1}{2}$. Using (3.15) for y and inverting the order of integration we find $A = 1/\sqrt{\pi}$ so that $C(0,t) = (t/\pi)^{1/2}$. This is not surprising because the source condition is equivalent to a steady flux of magnitude $\frac{1}{2}$ into each half-space.

7. The principal group is $C' = \lambda^\alpha C$, $t' = \lambda^{3/2} t$, $z' = \lambda z$. The second and third boundary conditions require $\alpha = 0$ so the most general invariant solution has the form $C = y(z/t^{2/3}) = y(x)$. The principal deq is $(\sqrt{x}\,\dot{y})^{\cdot} + \frac{2}{3} x\dot{y} = 0$ which must be solved with the boundary conditions $y(0) = 0$ and $y(\infty) = 1$. If we set $x = w^2$ we can separate variables, integrate, and find eventually that $y(x) = A \int_0^{\sqrt{x}} e^{-4w^3/9}\,dw$. Then we find $t^{1/3}\sqrt{z}\,C_z|_{z=0} = \sqrt{x}\,\dot{y}(x)|_{x=0} = A/2$, where A is determined by $1 = y(\infty) = A \int_0^\infty e^{-4w^3/9}\,dw$ to be $[(\frac{9}{4})^{1/3} \frac{1}{3} \Gamma(\frac{1}{3})]^{-1} = 0.855$.

8. The principal group is (4.2a) with the relation $(m-1)\alpha + \beta = 2$ between α and β. The source condition $\int_{-\infty}^{+\infty} C\,dz = 1$ obliges α to be -1 so that $\beta = m + 1$. Then $C = t^{-1/(m+1)} y(z/t^{1/(m+1)})$ is the most general form an invariant solution can take. The principal differential equation is $(y^m)^{\cdot\cdot} + (y + x\dot{y})/(m+1) = 0$, where $x = z/t^{1/(m+1)}$, and must be solved with the boundary conditions $\int_{-\infty}^{+\infty} y\,dx = 1$ and $y(\infty) = 0$. It can be integrated once directly to give $(y^m)^{\cdot} + xy/(m+1) = 0$ and then a second time after the variables have been separated to give $y = [\frac{m-1}{2m(m+1)}(x_0^2 - x^2)]^{1/(m-1)}$ for $0 < x < x_0$ and $y = 0$ for $x > x_0$. The constant x_0 is determined by the source condition $\int_{-\infty}^{+\infty} y\,dx = 1$.

9. Since the deq is invariant to the group $v' = \lambda v$, $u' = \lambda u$, we can use Lie's theorem, Section 2.2, to find that $[uv(2v - u)]^{-1}$ is an integrating factor. With it we can integrate (4.15) and obtain $v^4 u = \text{const}\,(2v - u)^3$. If $|v| \ll |u|$ when $|u| \gg 1$, we find that $v = \text{const} \times \sqrt{u}$, which is possible. If $|v| \gg |u|$ when $|u| \gg 1$, then $v = \text{const}/u$, which contradicts the hypothesis. If $v \sim u$, then u^5 and u^3 must be of the same order, which is impossible. So only $v \sim \sqrt{u}$ is possible.

10. Following the reasoning of Section 4.3 we find that the associated group is $y' = \mu y$, $x' = x$. If we introduce the invariant $u = x$ and the first

117

differential invariant $v = (\dot{y}/y)e^{x^2/4}$ we find the associated deq $\dot{v} = -v^2 e^{-x^2/4}$ when $\alpha = 0$. Then $v = (\frac{1}{v(0)} + \int_0^x e^{-x^2/4} dx)^{-1}$ so that $\dot{y}/y = e^{-x^2/4}/(\frac{1}{v(0)} + \int_0^x e^{-x^2/4} dx)$. Thus $\ln y = \ln(\frac{1}{v(0)} + \int_0^x e^{-x^2/4} dx) + \ln A$ or $y = A(\frac{1}{v(0)} + \int_0^x e^{-x^2/4} dx)$. Since $y(\infty) = 0$, $v(0) = -1/\sqrt{\pi}$. To make $y(0) = 1$, A must equal $v(0) = -1/\sqrt{\pi}$. Then after some easy rearrangement, we find $y = \text{erfc}(x/2)$.

11. Taking (4.2a) as the principal group, we find the linear constraint $n\alpha + \beta = 2$. The most general invariant solution has the form $C = t^{\alpha/\beta} y(z/t^{1/\beta})$, where acceptable functions $y(x)$ can be transformed into other acceptable functions by the transformations of the associated group $y' = \mu^{2/n} y$, $x' = \mu x$. The flux at the front face is given by $C^n C_z|_{z=0} = t^{[(n+1)\alpha-1]/\beta} \dot{y}(0) y^n(0)$. Images under the associated group all have the same value of the ratio $\dot{y}(0)/y^{(2-n)/2}(0)$; call it A. Then $C^n C_z|_{z=0} = At^{[(n+1)\alpha-1]/\beta} y^{(n+2)/2}(0) = At^{[(n+1)\alpha-1]/\beta}(C(0,t)t^{-\alpha/\beta})^{(n+2)/2}$. Collecting all the powers of t and using the linear constraint $n\alpha + \beta = 2$, we find that the net exponent of t is $-1/2$.

12. The principal group is $p' = \lambda^\alpha p$, $t' = \lambda^\beta t$, $z' = \lambda z$ with $\alpha - 2\beta = -3$. Now the most general invariant solution has the form $p = t^{\alpha/\beta} y(z/t^{1/\beta})$ and the principal ode for $y(x)$ is invariant to the associated group $y' = \mu^{-3} y$, $x' = \mu x$. Images under the associated group all have the same value of $\dot{y}(0)/y^{4/3}(0)$; call it $-A$. Then $-p_z|_{z=0} = -t^{(\alpha-1)/\beta} \dot{y}(0) = At^{(\alpha-1)/\beta}(p(0,t)t^{-\alpha/\beta})^{4/3} = At^{-2/3} p^{4/3}(0,t)$. Extracting a square root now gives the desired result.

13. The principal group is $C' = C + \alpha \ln \lambda$, $t' = \lambda^\beta t$, $z' = \lambda z$ where $\alpha + \beta = 2$. The most general invariant solution can be written as $C = (\alpha_0/\beta_0) \ln t + y(z/t^{1/\beta_0})$. Transforming this with members of the group for which $\alpha \neq \alpha_0$ and $\beta \neq \beta_0$, we find that $C = (\alpha_0/\beta_0) \ln t + y(\mu z/t^{1/\beta_0}) - 2 \ln \mu$, where $\mu = \lambda^{(\beta_0-\beta)/\beta_0}$, is also a solution. Finally, $y(\mu x) - 2 \ln \mu$ is the image of $y(x)$ under the group of transformations $y' = y + 2 \ln \mu$, $x' = \mu x$, which is the associated group. Since the principal ode for $y(x)$ is of second order, a solution is uniquely determined by two boundary values, say, $y(0)$ and $\dot{y}(0)$. So $y(\infty) = F(y(0), \dot{y}(0))$ and this relation must be invariant to the associated

group. By a now familiar procedure, we find that the most general invariant relation among these variables is $y(0) - y(\infty) = G(\dot{y}^2(0)e^{y(0)})$, where G is an arbitrary function.

14. All the terms in eq. (5.1a) are proportional to $\lambda^{2\alpha-\beta}$ under transformation by the group (5.2). For the term UU_x to transform in the same way, it must be proportional to the $(2\alpha - \beta)/\beta$ power of x, i.e., U must be proportional to a power of x. If $U \sim x^m$, then $2\alpha - \beta = (2m - 1)\beta$ so $\alpha = m\beta$. Together with the linear constraint (5.2b), this last relation determines α and β uniquely: $\alpha = -2m/(m - 1)$, $\beta = -2/(m - 1)$. Since now α and β are uniquely determined by the pde, the latter is invariant only to a single group of transformations and not to a one-parameter family of groups. So no associated group exists.

15. The temperature rise in the fluid is given by $T = \sqrt{x}\, g(\eta)$ where $\eta = y/\sqrt{x}$. The principal ode for g is $2D\ddot{g} = f\dot{g} - \dot{g}f$ where $f(\eta)$ is Blasius's function. The boundary conditions are $\dot{g}(0) = -q/D$, where q is the clamped wall flux, and $g(\infty) = 0$. (Remember to use the same special units as used in Section 5.5!) The problem is a two-point boundary value problem. For very large η the principal ode becomes $2D\ddot{g} = g - (\eta - a)\dot{g}$ where $a = 1.7208$. This deq is linear and has $\eta - a$ as a special solution. Using the method of Section 3.1 [cf. eqs. (3.13)-(3.15)] and the boundary condition $g(\infty) = 0$, we eventually find $g = A(\eta - a) \int_\eta^\infty e^{-(\eta-a)^2/4D} (\eta - a)^{-2}\, d\eta$, $\eta \gg 1$, where A is a constant of integration. From this expression we can obtain consistent values of g and $\dot{g}$ for some large value of η. Then by integrating inwards we can find $\dot{g}(0)$. Since the pde (5.35) and the principal ode above are linear, we can scale $\dot{g}(0)$ to the correct value by simply multiplying $g(x)$ by the appropriate factor without disturbing the condition that $g(\infty) = 0$.

16. $x\, du/dx = Y/x^{\alpha-1} - \alpha u$ and $x\, dv/dx = Z/x^{\beta-1} - \beta v$ so, if $Y/x^{\alpha-1}$ and $Z/x^{\beta-1}$ are functions only of u and v, du/dv will be, too. Invariance of the deqs to the group of transformations means that $\lambda^{\alpha-1}Y = Y(\lambda x, \lambda^\alpha y, \lambda^\beta z)$ from which it follows that $Y/x^{\alpha-1} = G(y/x^\alpha, z/x^\beta) = G(u,v)$. A similar conclusion holds for $Z/x^{\beta-1}$.

17. The strain is η on the positive characteristic with constant propagation velocity $c = \eta^{-1/4}$. So at time t the strain is η at the location $a = t\eta^{-1/4}$.

Thus $\eta = (a/t)^{-4}(\frac{a}{t})$. Since $v = -\frac{4}{3}\eta^{3/4}$ everywhere, $v = -\frac{4}{3}(\frac{a}{t})^{-3}$. This is precisely the exceptional solution mentioned in Section 6.4.

18. Suppose the stretching group is $v' = \lambda^\alpha v$, $\eta' = \lambda^\gamma \eta$, $t' = \lambda^\beta t$, and $a' = \lambda a$. Eq. (6.3) requires $\alpha - 1 = \gamma - \beta$; eq. (6.4b) requires $\sigma \sim \eta^m$ if it is to be invariant to the stretching group, in which case $\alpha - \beta = m\gamma - 1$. The shock conditions can be written $\rho_0 v^2 = \sigma\eta$ and $\rho_0 U^2 = \sigma/\eta$. The first condition will be invariant if $2\alpha = (m + 1)\gamma$, which follows from the two linear constraints given above. Now if the shock front is given by a constant value A of the similarity variable $a/t^{1/\beta}$, then $U = da/dt = (A/\beta)t^{(1-\beta)/\beta}$. Then $U' = \lambda^{1-\beta}U$ and the second shock condition will be invariant if $2(1 -\beta) = (m - 1)\gamma$, which also follows from the earlier linear constraints.

19. Following the first part of Section 6.8 we come to $u_{tt} = u_{zz}[\int_0^Z u_z^2 \, dz]^m$. This integro-pde is invariant to the group (6.48a-d) but with the constraint $2m\alpha + 2\beta = 2 + m$ instead of (6.48e). The distance Z advanced by the wave front at time t is an invariant function of p and t. Since Z scales as λ, t as λ^β, and p as $\lambda^{\alpha-2\beta} = \lambda^{[(2+m)/2m - (2m+1)\beta/m]}$ (cf. eq. (6.48f)), we can use the method of Section 4.15 to show that $Z \sim p^{2m/(2+m)} t^{(4m+2)/(m+2)}$. The time we seek is the time at which $Z = 1/2$, so it must scale as $p^{-m/(2m+1)}$.

20. If we set $T = t^{1/2}y(x,t)$ where $x = z/t^{1/2}$, we get for y the pde $2(y_x^{1/3})_x + xy_x - y = 2ty_t$. Now we set $y = p(t) y_1(r(t)x)$ where $y_1(x)$ is the similarity solution corresponding to a fixed value of q/k equal to the value of q(t)/k at time t. Then $y_x = pr \dot{y}_1$ so that $pr = 1$ (for $y_x(0,t) = \dot{y}_1(0) = -q^3(t)/k^3$). If we integrate the pde for y(x,t) over all x we get eq. (7.30a) with the left-hand side replaced by q(t)/k. This can be solved to give $p/r = \int_0^t q(t') \, dt'/t\, q(t)$ as in Section 7.3. Finally, then, $T(0,t) = a^{2/3} (q(t)/k)^2 t^{1/2}[\int_0^t q(t') \, dt'/t\, q(t)]^{1/2}$.

References

[AB68] "Handbook of Mathematical Functions," M. Abramowitz and I. A. Stegun, Dover Publications, New York, 1968.

[BA72] G. I. Barenblatt and Ya. B. Zeldovich, Annual Reviews of Fluid Mechanics 4: 285-312, 1972.

[BA79] "Similarity, Self-Similarity, and Intermediate Asymptotics," G. I. Barenblatt, N. Stein, translator, M. van Dyke, translation editor, Consultants Bureau, New York, 1979.

[BI50] "Hydrodynamics," G. Birkhoff, Princeton University Press, Princeton, New Jersey, Ch. V, 1950.

[BO94] L. Boltzmann, Ann. Physik (NF) 53:959, 1894.

[CO31] "An Introduction to the Lie Theory of One-Parameter Groups," A. Cohen, G. E. Stechert and Co., New York, 1931.

[CR75] "The Mathematics of Diffusion," J. Crank, Clarendon Press, Oxford, second edition, 1975.

[DA60] "Introduction to Nonlinear Differential and Integral Equations," H.T. Davis, U.S. Gov't. Printing Office, Wash., D.C., 1960.

[DI23] "Differential Equations from the Group Standpoint," L. E. Dickson, Annals of Mathematics, Second Series, vol. 25, pp. 287-378, 1923.

[DR64] "Boundary Layer Buildup in the Demineralization of Salt Water by Reverse Osmosis," L. Dresner, Oak Ridge National Laboratory Report ORNL/TM-3621, May, 1964.

[DR70] L. Dresner, J. Appl. Phys. 41: 2542, 1970.

[DR71] L. Dresner, J. Math. Phys. 12 (7): 1339, 1971.

[DR80] "On the Calculation of Similarity Solutions of Partial Differential Equations," L. Dresner, Oak Ridge National Laboratory Report ORNL/TM-7404, Aug., 1980.

[DR81] "Thermal Expulsion of Helium from a Quenching Cable-in-Conduit Conductor," L. Dresner, Proc. of the Ninth Symposium on the Engineering Problems of Fusion Research, Chicago, IL, Oct. 26-29, 1981, IEEE Publication No. 81CH1715-2NPS, pp. 618-621.

[DR82] L. Dresner, Advances in Cryogenic Engineering, R. W. Fast (ed.), vol. 27, pp. 411-419, Plenum Publishing Corp., 1982.
[KA50] T. von Karman and P. Duwez, J. Appl. Phys. $\underline{21}$: 987, 1950.
[LE62] "Physicochemical Hydrodynamics," V. G. Levich, Scripta Technica, 1962.
[MA69] B. F. Maddock, G. B. James and W. F. Norris, Cryogenics $\underline{9}$: 261, 1969.
[NE]   J. von Neumann in "Blast Waves," Los Alamos Scientific Laboratory Technical Series, LA-2000, vol. VII, pt. II, H. Bethe (ed.).
[NI81] R. H. Nilson, J. Fluids Engineering $\underline{103}$: 339, 1981.
[PA59] R. E. Pattle, Quart. J. Mech. and Appl. Math., vol. XII, part 4, pp. 407-409, 1959.
[SC79] S. W. van Sciver, Cryogenics $\underline{19}$: 385, 1979.
[SE59] "Similarity and Dimensional Methods in Mechanics," L. I. Sedov, Academic Press, New York, 1959.
[SH65] T. K. Sherwood, P. L. T. Brian, R. E. Fisher and L. Dresner, Ind. and Engr. Chem. Fund. $\underline{4}$: 113, 1965.
[SP70] E. M. Sparrow, H. Quack and C. J. Boerner, AIAA Journal $\underline{8}$: 1936-42, 1970.
[TA50] G. I. Taylor, Proc. Roy. Soc. $\underline{A201}$: 159, 1950.
[TA58] "Plastic Wave in a Wire Extended by an Impact Load," G. I. Taylor, Scientific Papers, vol. I, Cambridge University Press, New York, 1958.

GENERAL BIBLIOGRAPHY

Two excellent sources of information on the Lie theory of differential equations are references CO31 and DI23. Other books on the subject of similarity solutions are references BA79 and SE59 and the works cited below.

"Similarity Analysis of Boundary-Value Problems in Engineering," A. G. Hansen, Prentice-Hall, 1964.

"Similarity Methods for Differential Equations," G. W. Bluman and J. D. Cole, Springer-Verlag, 1974.

"Nonlinear Partial Differential Equations in Engineering," W. F. Ames, vol. II, Chap. 2, pp. 87-145, Academic Press, 1972.

Index

Birkhoff, Garrett 2, 17
Blasius's equation 57
Boltzmann, Ludwig 2, 25
Boundary conditions
  collapse of 19, 21
  two-point 4, 58, 68, 76
Boundary layer 55
  free convection 67
  thermal 65

Characteristic equations 12, 13, 18
Characteristics 77
Concentration polarization 21

Darcy's law 25
Differential equation
  associated 3, 59
  first-order ordinary 5
  ordinary 5
  principal 3
  second-order ordinary 12
Diffusion equation
  in cylindrical geometry 101
  nonlinear 25
  ordinary 17
Dimensional analysis 52
Direction field 3, 5, 8, 15, 28
  (see also, Lie plane)

Eigenvalue problems 111

Emden-Fowler equation 14
Exceptional solutions 35, 37, 45, 46, 47, 76

Gorter-Mellink law 41
Group
  associated 3, 31, 35, 51
  principal 3
  stretching 2, 31, 33
  translation 70, 97
Group invariant, see invariant

Integrating factor 6
Invariant 11, 12
  first differential 12
Invariant solutions 17

Karman-Duwez-Taylor problems 69, 70

Lie plane 3, 4, 37, 43, 51, 76, 85, 99
  (see also, Direction field)
Lie's theorems 1, 6, 12
Lie, Sophus 1

Numerical integration 4, 29, 39

Porous medium, percolation in 25, 33, 47

Riemann invariants 77

Separation of variables 11

Separatrices 4, 10, 38, 45, 76, 87
Shock conditions 81
Shock front 81
Shock-loaded membrane 87
Similarity, local 64, 107
Similarity solution 2, 3, 21, 104
Similarity variables 19
Singular points 4, 9
Special units 17

Thermal expulsion of fluid 35

Wave propagation 69
Waves
  in a channel 95
  travelling 97